Molecular
Biology
Intelligence
Unit

Tumor Suppressors

Involvement in Human Diseases, Viral Protein Interactions and Growth Regulation

John W. Ludlow, Ph.D.
Gary R. Skuse, Ph.D.

Departments of Biochemistry and Medicine
University of Rochester
Rochester, New York

R.G. Landes Company
Austin

MOLECULAR BIOLOGY INTELLIGENCE UNIT

TUMOR SUPPRESSORS: INVOLVEMENT IN HUMAN DISEASES, VIRAL PROTEIN INTERACTIONS AND GROWTH REGULATION

R.G. LANDES COMPANY
Austin

CRC Press is the exclusive worldwide distributor of publications of the Molecular Biology Intelligence Unit.
CRC Press, 2000 Corporate Blvd., NW, Boca Raton, FL 33431. Phone: 407/994-0555.

Submitted: June 1994
Published: September 1994

Production Manager: Debbie Molsberry
Production Associate: Mary Edwards
Copy Editor: Constance Kerkaporta

Please address all inquiries to the Publisher:
R.G. Landes Company, 909 Pine Street, Georgetown, TX 78626
or
P.O. Box 4858, Austin, TX 78765

Phone: 512/ 863 7762; FAX: 512/ 863 0081

ISBN 1-57059-105-9
CATALOG # LN9105

Library of Congress Cataloging-in-Publication Data

Ludlow, John W.

Tumor suppressors: involvement in human diseases, viral protein interactions and growth regulation / John W. Ludlow. Gary R. Skuse.

p. cm.—(Molecular biology intelligence unit)

Includes bibliographical references and index.

ISBN 1-57059-105-9

1. Antioncogenes. 2. Oncogenesis. 3. Oncogenic viruses. 4. Cell proliferation. 5. Cancer cells—Growth—Regulation. 6. Cells—Growth—Regulation. I. Skuse, Gary R. II. Title. III. Series.

[DNLM: 1. Oncogenes—physiology. 2. Genes, Suppressor, Tumor—physiology. 3. Oncogene Proteins, Viral—metabolism. 4. Growth Substances—physiology. 5. Cell Transformation, Neoplastic. QH 447 L945t 1994]

RC268.43.L83 1994
616.99'2071—dc20
DNLM/DLC
for Library of Congress 94-3687
CIP

To Our Readers

R.G. Landes Company publishes four book series: *Medical Intelligence Unit, Molecular Biology Intelligence Unit, Neuroscience Intelligence Unit,* and *Biotechnology Intelligence Unit.* Our goal is to publish the most recent information in biomedical science for sophisticated researchers and physicians.

To achieve this goal we have accelerated our publishing program to conform to the fast pace in which information grows in biomedical science. The book you have in hand, like all titles in our series, was published *within 90 days of receipt of the manuscript.*

As you might expect, this causes a few problems for us; sometimes it makes our office more like a big city newspaper than a scholarly publisher. So as you look through this book you may see something that isn't just right. Please let us know. Or if you have an idea for improving our books, we'd like very much to hear from you. If the problem you describe hasn't already been discovered or if your idea is provocative enough to promote a discussion here, we'll give you a free book of your choice. Just list three titles in order of preference and we'll send you one, based on availability—with our thanks. Our address is printed on the copyright page of each of our books.

Elinda McKenna
Director of Operations
R.G. Landes Company

Acknowledgments

We would like to acknowledge the grant support of the National Cancer Institute to us as individuals (CA56940 awarded to J.W.L.; CA55173 awarded to G.R.S.) and to the University of Rochester Cancer Center (CA11198 awarded to the late Robert A. Cooper Jr.). We dedicate this monograph to our ever tolerant partners, MaryBeth and Jan and to Nick, Megan, and Kristen.

John W. Ludlow

Gary. R. Skuse

June 1994

CONTENTS

PREFACE

We have strived to accomplish two goals by this writing. The first is to provide an overview of the field to those readers newly acquainted with tumor suppressors. Towards this goal, we have included what we hope are simpler versions of what may be considered by some to be complex diagrams and models. Additionally, a summary paragraph is included at the end of most chapters highlighting its central theme. Our second goal was to also include sufficient detail, both experimentally and in literature citations, for those readers who are more familiar with this field. In this way, perhaps new ideas or directions of research in the areas of tumor suppressor function, cell cycle regulation, and viral oncoprotein function may be fostered. Furthermore we invite the reader to think about clinical applications which may be based on our current understanding of tumor suppressor genes.

While every effort was made on our part to include the most current and pertinent information gleaned from the published literature, in such a rapidly expanding and changing field, it is obviously not possible to include each and every published report. Along this same line, inclusion of all of the past literature leading the way to present hypotheses and models is beyond the scope of this writing. With this said, we sincerely apologize to those of our colleagues whose work has been omitted due to our oversight or lack of space.

For those readers who desire more detailed literature in some of the specific areas covered in this monograph, we suggest beginning your search in the recent reviews by Wiman,[1] Zambetti and Levine,[2] Ludlow,[3] Vousden,[4] Moran,[5] and Coppes et al,[6] information from which is summarized throughout this work.

References

1. Wiman KG. The retinoblastoma gene: role in cell cycle control and differentiation. FASEB J 1993; 7:841-845.
2. Zambetti GP, Levine AJ. A comparison of the biological activities of wild-type and mutant p53. FASEB J 1993; 7:855-865.
3. Ludlow, JW. Interactions between SV40 large-tumor antigen and the growth suppressor proteins pRB and p53. FASEB J 1993; 7:866-871.
4. Vousden K. Interactions of human papillomavirus transforming proteins with the products of tumor suppressor genes. FASEB J 1993; 7:872-879.
5. Moran E. Interaction of adenoviral proteins with pRB and p53. FASEB J 1993; 7:880-885.
6. Coppes MJ, Campbell CE, Williams BRG. The role of *WT1* in Wilms tumorigenesis. FASEB J 1993; 7:886-895.

CHAPTER 1

A Historical Perspective

A. DISCOVERY OF ONCOGENES

In order to gain an appreciation of the current state of our knowledge of tumor suppressor genes, it is necessary to first review what is known about oncogenes. Simply stated, oncogenes and tumor suppressor genes are two sides of the same coin. In a normal or untransformed cell, oncogene and tumor suppressor gene products function to regulate cell proliferation. Once activated, however, oncogenes function in a dominant manner while tumor suppressor genes act recessively. Thus, only one copy of a mutated or activated oncogene is sufficient to give rise to malignant transformation while both alleles of a tumor suppressor gene must typically be inactivated for a tumor to develop.

The existence of oncogenes was first revealed when Rous discovered that a virus was the etiological agent responsible for sarcoma development in chickens.[1] Further investigation revealed that the particular gene responsible for tumor initiation carried by the Rous sarcoma virus had a cellular homolog which did not normally promote unregulated growth.[2,3] The cellular homolog, termed a protoocogene because of its oncogenic potential and written with a preceding letter "c" to designate it as the cellular homolog, did play a role in growth regulation which was undermined when the gene was transduced by an infecting virus.[4]

B. PROPERTIES OF ONCOGENE PRODUCTS

The products of protooncogenes fall into categories of proteins with related functions (summarized in Table 1.1). It is interesting to note that, like the known tumor suppressor gene products, the products of oncogenes are distributed throughout the cell. One category of oncogene product, the protein kinases, exemplified by *src* includes a family of tyrosine or threonine kinases which can be found in the plasma membrane or in the cytoplasm. Another category includes the three members of the *ras* family of guanine nucleotide binding proteins which are found associated with the plasma membrane. A third class of oncoproteins includes the nuclear proteins which regulate transcription and is exemplified by *myc*, *myb*, *fos*, and others. Still other oncogene products function as growth factors or growth factor receptors such as the proteins encoded by *ros* and *sis*. Together the products of oncogenes act at a variety of levels within the cell to regulate cellular proliferation and differentiation. In view of their respective normal functions, the mechanisms behind their role in malignant transformation

Table 1.1. A partial list of the known oncogenes

Protooncogene	Associated Retrovirus	Subcellular Location of Gene Product	Chromosomal Location of Gene	Function
abl	Abelson murine leukemia virus	plasma membrane	9q	tyrosine kinase
erb-A	avian erythroblastosis virus	cytoplasm	17q	growth factor receptor
erb-B	avian erythroblastosis virus	plasma membrane	7p	tyrosine kinase
fes	feline sarcoma virus	plasma membrane	15q	tyrosine kinase
fgr	Gardner-Rasheed sarcoma virus	plasma membrane		tyrosine kinase
fos		nucleus		transcription factor
mos	Moloney sarcoma virus	cytoplasm	8q	serine/threonine kinase
raf1	murine sarcoma virus	cytoplasm	3p	
H-ras	Harvey sarcoma virus	cytoplasm	11p	signal transduction
K-ras	Kirsten sarcoma virus	cytoplasm	12p	signal transduction
fms	McDonough sarcoma virus	plasma membrane	5q	tyrosine kinase
sis	Simian sarcoma virus	extracellular	22q	platelet derived growth factor B
myc	MC29 myelocytomatosis virus	nucleus	8q	transcription factor
myb	avial myeloblastosis virus	nucleus	6q	transcription factor
rcs	avian leukosis virus	plasma membrane		growth factor receptor
erbB-2		plasma membrane	17q	estrogen receptor
N-myc	none	nucleus	1p	transcription factor

becomes apparent. When their function in regulating cellular proliferation is perturbed cell growth most often becomes uncontrolled thus facilitating tumor development.

C. ONCOGENES IN HUMAN DISEASE

The oncogene *ras* was first identified in cultured EJ bladder carcinoma cells when DNA from those cells was transfected onto NIH 3T3 mouse fibroblasts giving rise to foci of malignantly transformed cells.[5] DNA was isolated from those foci and used in serial transfections in order to isolate the DNA carrying the transforming gene. This led to the identification of the activated *ras* oncogene as the agent responsible for cellular transformation. Since its identification in EJ cells several viral and cellular homologs of *ras* have been identified. *Ras* is frequently mutated by environmental factors such as chemical carcinogens and ionizing radiation so that cellular transformation occurs without viral involvement.[6]

Protooncogenes may become activated by mutation, as is often the case with *ras*, or by several mechanisms which will lead to their inappropriate expression. For example this can occur by the integration of an otherwise non-oncogenic retrovirus near the promoter of a growth regulating protooncogene. This leads to increased levels of protooncogene transcription due to the proximity of an active viral promoter. Similar events occur without viral intervention in several human diseases such as Burkitt's lymphoma.[7,8] Here, a region of chromosome 8 carrying the c-*myc* gene is translocated to chromosome 14, an actively transcribed region which carries immunoglobulin loci resulting in over expression of the *myc* protooncogene and chronic myelogenous leukemia.[9]

A consistent chromosomal aberration occurring in chronic myelogenous leukemia (CML) involves a reciprocal translocation between chromosomes 9 and 22.[10] In that process the *abl* protooncogene, first identified in the Abelson murine leukemia virus,[11-14] is juxtaposed to the 5' portion the *bcr* gene of chromosome 22.[12,15,16] This translocation results in a truncation of *abl* in which a 5' portion of the gene is lost. A similar 5' truncation results in the oncogenic activation of *abl* in the Abelson murine leukemia virus. Interestingly another protooncogene, *sis*, is translocated from chromosome 22 to chromosome 9 in CML but does not appear to be involved in the disease phenotype.

D. ONCOGENE SUMMARY

Since the identification of *src*, the oncogene present in the Rous sarcoma virus, many more oncogenes and protooncogenes have been identified. A partial list of the more than 100 oncogenes identified to date is presented in Table 1.1. Each of them plays a fundamental role in regulating cell proliferation and can cause unregulated growth when activated by mutation. Oncogenes typically act in a dominant fashion at the cellular level in that only one copy of a protooncogene needs to become activated for tumor initiation to occur. In this way an oncogene introduced by an infecting virus can transform a cell in the presence of the normal cellular homologs and a single mutation in a cellular oncogene can result in malignant transformation of that cell.

E. DISCOVERY OF TUMOR SUPPRESSOR GENES

In a review by Touchette,[17] information from which is summarized below, we are reminded that the discovery of human tumor suppressor genes was greatly facilitated by earlier studies carried out in lower organisms. For example, Gateff and Schneiderman[18] determined during the late 1960s that mutations at the lethal(2)giant larvae (l(2)gl) locus in *Drosophila* resulted in tumor formation in precursor cells of the optic centers of adult flies. This was one of the first indications that a mutation in a recessive gene may lead to neoplastic growth. During this same time period, Henry Harris and George Klein[19] demonstrated that fused normal rodent cells and tumor cells were no longer cancerous. Following this observation, Stanbridge[20] formed stable non-tumorigenic hybrid cells from normal and tumor derived human cells. Taken together, these studies suggested that a component of the normal cell could suppress malignant growth thus supporting the notion that a new class of genes capable of suppressing or negatively regulating the growth of tumors may exist.

Utilizing the powerful techniques of molecular biology, Mechler cloned the first tumor suppressor gene, the *l(2)gl* gene from *Drosophila*, in 1985.[21] Shortly thereafter, Stanbridge demonstrated that the transfer of chromosome 11 could suppress the malignant phenotype of HeLa cells.[22] Later that same year the human retinoblastoma gene, *RB1*, was cloned by Thaddeus Dryja, Stephen Friend, Robert Weinberg and their colleagues.[23] *RB1* is now considered the prototype tumor suppressor gene. To demonstrate the tumor suppressor function of the *RB1* gene Wen-Hwa Lee and colleagues transferred this gene into retinoblastoma and osteosarcoma cells and found that cells expressing this gene product, designated pRB, were no longer tumorigenic.[24] To date, this is the only direct demonstration that a tumor suppressor can suppress tumor formation.

In 1989, Arnold Levine and his colleagues proposed that *p53* is a tumor suppressor gene.[25] Originally, *p53* was thought to be an oncogene by virtue of the fact that this gene can immortalize cells on its own or cooperate with the *ras* oncogene to transform cells. It was discovered, however, that the transforming *p53* gene originally identified carried a mutation. Subsequently, the wild-type *p53* was shown to be capable of inhibiting or suppressing the ability of *ras* to transform cells. Thus, the product of this gene, designated p53, can function in two diametrically opposed roles—as a tumor suppressor in the wild-type or unmutated form, and as an oncoprotein that directly promotes tumor growth in the mutant form. As the dual role of p53 was being revealed different groups led by Samuel Benchimol[26] and Bert Vogelstein[27] independently discovered that both copies of *p53* were altered in certain types of cancer tissue. *p53* is atypical among tumor suppressor genes in that one mutated allele will act in a dominant negative fashion to give cells heterozygous for that mutation a selective growth advantage. The cells with that advantage divide more often than the ones without it and therefore have a greater probability of acquiring a second *p53* mutation and thus becoming tumorigenic.

F. TUMOR SUPPRESSOR GENE INACTIVATION

One difficulty encountered during the experimental study of tumor suppressor genes is that their existence becomes most evident when they are absent from the cell genome. As one could imagine being faced with such a dilemma, progress in this area of research has occurred much more slowly than similar work focused on oncogene action. In attempting to develop a model to explain tumor suppressor gene inactivation, Alfred Knudson in 1971 put forth the notion that retinoblastoma, a rare form of childhood eye cancer, arose as a consequence of two successive alterations in the cell genome.[28] This disease can be found in both familial, or inherited, form as well as a sporadic, spontaneously arising form. Therefore, the model had to be inclusive of both possible scenarios. For the familial or inherited form of retinoblastoma, Knudson proposed that one of the two mutations is acquired from a genetically afflicted parent. Alternatively, he proposed that this mutation could also occur during gametogenesis. The model proposes that a second genetic mutation then occurs as a somatic or postconceptional event. In the sporadic form of this disease, the afflicted individual inherits two functional copies of this gene and both copies in the retinal cell precursors acquire mutations somatically. This model has become known as the Knudson "two-hit hypothesis" because of the need for two inactivating mutations which lead to tumor formation. Implicit in this model is the notion that functional inactivation of both copies of *RB1* is required for clinical manifestation of retinoblastoma; one intact copy of this gene is sufficient to protect against this disease. It was not until the 1980s, however, that the genetic targets Knudson built this model on were localized to the two copies of *RB1* on chromosome 13q14 and the two mutational events which he proposed involved the inactivation of both functional copies of this gene.

G. PROSPECTS FOR TUMOR SUPPRESSOR GENES

Since the molecular cloning of *RB1*, and the discovery that viral tumor oncoproteins can form specific complexes with the product of this gene,[29,30] much attention has been given to the function of tumor suppressor genes and their protein products. With such intensity, it is little wonder that research has revealed new tumor suppressor genes with increasing frequency (Table 1.2, timetable of tumor suppressor milestones). In 1990, the Wilms' tumor gene, a candidate tumor suppressor gene believed to be involved in childhood kidney cancer, was cloned and sequenced independently by groups led by David Housman[31] and Gail Bruns.[32] Also in 1990, Francis Collins and colleagues[33] and, independently, Ray White and colleagues[34] cloned the neurofibromatosis gene (*NF1*). Mutations in one copy of this gene results in neurofibromatosis, and defects in both copies may lead to peripheral neurofibrosarcomas and an increased risk of other malignancies. Vogelstein and his associates in 1990 found another type of tumor suppressor gene located on the q21 band of chromosome 18.[35] That gene, named *DCC* because it is deleted in colon carcinomas, codes for a 1000+ amino acid protein and shares a large part of its sequence with cell-adhesion molecules involved in cell-cell communication.

H. TUMOR SUPPRESSOR GENE SUMMARY

Since the mid-1980s, researchers in the area of cancer biology have been refining their knowledge of a quite different class of cancer genes that act not to promote cellular proliferation, but to suppress it (Table 1.3, list of known tumor suppressor genes). By eliminating the function of genes that normally act to restrain cell proliferation, cells may proliferate uncontrollably thus increasing the probability that the cell will become cancerous. Although not often implicitly stated, many hope that by understanding how the products of these genes function to suppress cell growth, significant contributions can be made to the clinical arena in treating neoplastic diseases, perhaps by stimulating the body's own natural mechanisms to halt cancer cell growth.

Table 1.2. Milestones in tumor suppressor gene research

Year	Achievement	Reference
1967	Observed that mutations of a specific gene in *Drosophila* result in tumors of imaginal discs	18
1969	Found that hybrid cells resulting from fusion of normal and tumor rodent cells do not cause tumors	19
1976	Stable non-tumorigenic hybrid cells were formed from normal and tumorous human cells	20
1985	Cloning of the first tumor suppressor gene from *Drosophila*	21
1986	Demonstration that transfer of specific chromosome could suppress malignant phenotype in cultured cells	22,37
1986	Cloning of the retinoblastoma susceptibility gene	23
1989	p53 reclassified as a tumor suppressor	25
1989	NF1 gene demonstrated as a tumor suppressor	36
1990	Wilms' tumor gene cloned	31,32
1990	NF1 gene cloned	33,34
1990	DCC gene localized to chromosome 18q and cloned	35

Compiled from Touchette.[17]

Table 1.3. A partial list of known tumor suppressor genes

Disease or Tumor Type	Chromosomal Location	Has the Gene Been Identified?	References
MEN 1	11	No	45
MEN 2a	10	No	42
Neuroblastoma	1p	No	46,47
Melanoma	9p	No	48,49
Renal Cell Carcinoma	3p	No	38
von Hippel Lindau Disease	3p	Yes	39
Small Cell Lung Cancer	13p, 9	No	43,44
Familial Polyposis coli	5q	Yes	50,51
Colon Carcinoma	5q, 17p, 18q	Yes	35
Wilms' Tumor	11p	Yes	31,32
Beckwith-Wiedemann Syndrome	11p	No	52
Retinoblastoma	13q	Yes	23
Astrocytoma	17p	No	40
Neurofibromatosis-1	17q	Yes	33,34
Neurofibromatosis-2	22q	Yes	41
Meningioma	22q,10	No	53,54

References

1. Rous PA. Sarcoma of the fowl transmissible by an agent separable from the tumor cells. J Exp Med 1911; 13:397-411.
2. Oppermann H, Levinson AD, Varmus HE et al. Uninfected vertebrate cells contain a protein that is closely related to the product of the avian sarcoma virus transforming gene (*src*). Proc Natl Acad Sci USA 1979; 76:1804-1808.
3. Takeya T, Hanafusa H. Structure and sequence of the cellular gene homologous to the RSV *src* gene and the mechanism for generating the transforming virus. Cell 1983; 32:881-890.
4. Temin H. Nature of the provirus of Rous sarcoma. Natl Cancer Inst Monogr 1964; 17:557-570.
5. Shih C, Shilo B-Z, Goldfarb MP et al. Passage of phenotypes of chemically transformed cells via transfection of DNA and chromatin. Proc Natl Acad Sci USA 1979; 76:5714-5718.
6. Sukumar S, Notario V, Martin-Zanca D et al. Induction of mammary carcinomas in rats by nitroso-methylurea involves malignant activation of H-*ras*-1 locus by single point mutations. Nature 1983; 306:658-661.
7. Dalla-Favera F, Martinotti S, Gallo R et al. Translocation and rearrangements of the c-*myc* oncogene locus in human undifferentiated B-cell lymphomas. Science 1983; 219:963-967.
8. Taub R, Kirsh I, Morton C et al. Translocation of the c-*myc* gene into the immunoglobulin heavy chain locus in human Burkitt lymphoma and murine plasmacytoma cells. Proc Natl Acad Sci USA 1982; 79:7837-7841.
9. Rowley JD. A new consistent chromosomal abnormality in chronic myelogenous leukemia. Nature 1973; 243:290-293.
10. Sanberg AA. Chromosomes and causation of human cancer and leukemia. The Ph1 and other translocations in CML. Cancer 1980; 46:2221-2226.
11. Goff SP, Gilboa E, Witte ON et al. Structure of the Abelson murine leukemia genome and the homologous cellular gene: studies with cloned viral DNA. Cell 1980; 22:777-785.
12. Heisterkamp N, Stephenson JR, Groffen J et al. Localization of the c-*abl* oncogene adjacent to a translocation breakpoint in chronic myelocytic leukemia. Nature 1983; 306:239-242.
13. Wang JTJ, Ledley F, Goff S et al. The mouse c-*abl* locus. Molecular cloning and characterization. Cell 1984; 36:349-356.
14. Witte ON, Rosenberg N, Baltimore D. A normal cell protein cross-reactive to the major Abelson murine leukemia virus gene product. Nature 1979; 281:396-398.
15. Bartram CR, de Klein A, Hagemeijer A et al. Translocations of the c-*abl* oncogene correlates with the presence of a Philadelphia chromosome in chronic myelocytic leukemia. Nature 1983; 306:277-280.
16. Groffen J, Stephenson JR, Heisterkamp N et al. Philadelphia chromosomal breakpoints are clustered within a limited region, *bcr*, of chromosome 22. Cell 1984; 36:93-99.
17. Touchette N. Tumor suppressors: a new arena in the war against cancer. J NIH Res 1990; 2:62-66.

18. Gateff E, Schneiderman HA. Developmental studies of a new mutation of *Drosophila melanogaster*: lethal malignant brain tumor l(2)gl. Amer Zoo 1967; 7:760.
19. Harris H, Miller OJ, Klein G et al. Suppression of malignancy by cell fusion. Nature 1969; 223:363-368.
20. Stanbridge EJ. Suppression of malignancy in human cells. Nature 1976; 260:17-20.
21. Mechler BM, McGinnis W, Gehring WJ. Molecular cloning of *lethal(2)giant larvae*, a recessive oncogene in *Drosophila melanogaster*. EMBO J 1985; 4:1551-1557.
22. Saxon PJ, Srivatsan ES, Stanbridge EJ. Introduction of chromosome 11 via microcell transfer controls tumorigenic expression of HeLa cells. EMBO J 1986; 5:3461-3466.
23. Friend SH, Bernards R, Rogelj S et al. A human DNA segment with properties of the gene that predisposes to retinoblastoma and osteosarcoma. Nature 1986; 323:643-646.
24. Huang H-JS, Yee J-K, Shew J-Y et al. Suppression of the neoplastic phenotype by replacement of the RB gene in human cancer cells. Science 1988; 242:1563-1566.
25. Finlay CA, Hinds PW, Levine AJ. The p53 proto-oncogene can act as a suppressor of transformation. Cell 1989; 57:1083-1093.
26. Benchimol S, Munroe DG, Peacock J et al. Abnormalities in structure and expression of the p53 gene in leukemia. Cancer Cells 1989; 7:121-125.
27. Nigro JM, Baker SJ, Preisinger AC et al. Mutations in the p53 gene occur in diverse human tumor types. Nature 1989; 342:705-708.
28. Knudson AG. Mutation and cancer: Statistical study of retinoblastoma. Proc Natl Acad Sci USA 1971; 68:820-823.
29. Whyte P, Buchkovich KJ, Horowitz JM et al. Association between an oncogene and an antioncogene: the adenovirus E1A proteins bind to the retinoblastoma gene product. Nature 1988; 334:124-129.
30. DeCaprio JA, Ludlow JW, Figge J et al. SV40 large tumor antigen forms a specific complex with the product of the retinoblastoma susceptibility gene. Cell 1988; 54:57-65.
31. Call KM, Glasser T, Ito CY et al. Isolation and characterization of a zinc finger polypeptide gene at the human chromosome 11 Wilms' tumor locus. Cell 1990; 60:509-520.
32. Gessler M, Poustka A, Cavenee W et al. Homozygous deletion in Wilms' tumours of a zinc-finger gene identified by chromosome jumping. Nature 1990; 343:774-778.
33. Wallace MR, Marchuk DA, Andersen LB et al. Type 1 neurofibromatosis gene: Identification of a large transcript disrupted in three NF1 patients. Science 1990; 249:181-186.
34. Viskochil D, Buchberg AM, Xu G et al. Deletions and a translocation interrupt a cloned gene at the neurofibromatosis type 1 locus. Cell 1990; 62:187-192.
35. Fearon ER, Cho KR, Nigro JM et al. Identification of a chromosome 18q gene that is altered in colorectal cancers. Science 1990; 247:49-56.
36. Skuse GR, Kosciolek BA, Rowley PT. Molecular genetic analysis

of tumors in von Recklinghausen neurofibromatosis: loss of heterozygosity for chromosome 17. Genes Chrom Cancer 1989; 1:36-41.
37. Weissman BE, Saxon PJ, Pasquale SR et al. Introduction of a normal human chromosome 11 into a Wilms' tumor cell line controls its tumorigenic expression. Science 1987; 236:175-180.
38. Erlandsson R, Boldog F, Persson B et al. The gene from the short arm of chromosome 3, at D3F15S2, frequently deleted in renal cell carcinoma, encodes acylpeptide hydrolase. Oncogene 1991; 6:1293-1295.
39. Latif F, Tory K, Gnarra J et al. Identification of the von Hippel-Lindau disease tumor suppressor gene. Science 1993; 260:1317-1320.
40. von Deimling A, Eibl RH, Ohgaki H et al. p53 mutations are associated with 17p allelic loss in grade II and grade III astrocytoma. Cancer Res 1992; 52:2987-2990.
41. Trofatter JA, MacCollin MM, Ruter JL et al. A novel, moesin-radixin-like gene is a candidate for the neurofibromatosis type 2 tumor suppressor. Cell 1993; 72:791-800.
42. Mole SE, Mulligan LM, Healey CS et al. Localisation of the gene for multiple endocrine neoplasia type 2A to a 480 kb region in chromosome band 10q11.2. Hum Mol Genet 1993; 2:247-252.
43. Daly MC, Xiang RH, Buchhagen D et al. A homozygous deletion on chromosome 3 in a small cell lung cancer cell line correlates with a region of tumor suppressor activity. Oncogene 1993; 8:1721-1729.
44. Olopade OI, Buchhagen DL, Malik K et al. Homozygous loss of the interferon genes defines the critical region on 9p that is deleted in lung cancers. Cancer Res 1993; 53:2410-2415.
45. Thakker RV, Woodling C, Pand JT et al. Linkage analysis of 7 polymorphic markers at chromosome 11p11.2-11q13 in 27 multiple endocrine neoplasia type 1 families. Annals Hum Genet 1993; 57:17-25.
46. Caron H, van Sluis P, van Hoeve M et al. Allelic loss of chromosome 1p36 in neuroblastoma is of preferential maternal origin and correlates with N-*myc* amplification. Nat Genet 1993; 4:187-190.
47. Stock C, Ambros IM, Mann G et al. Detection of 1p36 deletions in paraffin sections of neuroblastoma tissues. Genes Chrom Cancer 1993; 6:1-9.
48. Petty EM, Gibson LH, Fountain JW et al. Molecular definition of a chromosome 9p21 germ-line deletion in a woman with multiple melanomas and a plexiform neurofibroma: implications for 9p tumor-suppressor gene(s). Am J Hum Genet 1993; 53:96-104.
49. Coleman A, Fountain JW, Nobori T et al. Distinct deletions of chromosome 9p associated with melanoma versus glioma, lung cancer, and leukemia. Cancer Res 1994; 54:344-348.
50. Kinzler KW, Nilbert MC, Vogelstein B et al. Identification of a gene located at chromosome 5q21 that is mutated in colorectal cancers. Science 1991; 251:1366-1370.
51. Nakamura Y, Nishisho I, Kinzler KW et al. Mutations of the adenomatous polyposis coli gene in familial polyposis coli patients

and sporadic colorectal tumors. Int Symp Princess Takamatsu Cancer Res Fund 1991; 22:285-292.

52. Weksberg R, Teshima I, Williams BR et al. Molecular characterization of cytogenetic alterations associated with the Beckwith-Wiedemann syndrome (BWS) phenotype refines the localization and suggests the gene for BWS is imprinted. Hum Mol Genet 1993; 2:549-556.
53. Rempel SA, Schwechheimer K, Davis RL et al. Loss of heterozygosity for loci on chromosome 10 is associated with morphologically malignant meningioma progression. Cancer Res 1993; 53:2386-2392.
54. Xie YG, Han FY, Peyrard M et al. Cloning of a novel, anonymous gene from a megabase-range YAC and cosmid contig in the neurofibromatosis type 2/meningioma region on human chromosome 22q12. Hum Mol Genet 1993; 2:1361-1369.

CHAPTER 2

Tumor Suppressors in Human Disease

A. INTRODUCTION

As detailed in chapter one, the existence of a class of genes which has the ability to suppress tumor development in human disease was postulated by Knudson[1-3] based on his analysis of the incidence of retinoblastoma. He observed that retinoblastoma occurred in both a familial and a sporadic form. The familial form displayed an earlier age of onset compared to the sporadic disease and involved either bilateral or multicentric tumors (see Fig. 2.1). Those observations led to the hypothesis that a tumor suppressor may be involved in the pathogenesis of retinoblastoma and that two mutations must be incurred before a tumor will develop. This hypothesis is commonly known as the "two-hit hypothesis". Comings later extended that hypothesis to specify that the mutations occur in each of two alleles of the same tumor suppressor gene.[4]

An individual with the familial form of retinoblastoma is born with one allele already inactivated by mutation in every cell. A tumor arises when the second allele is inactivated by mutation in a susceptible cell, i.e. a retinoblast. The need for only one additional post-conceptional mutation accounts for the earlier age of onset and the occurrence of multiple tumors in individuals with familial retinoblastoma. An individual with sporadic retinoblastoma must incur two inactivating mutations, one in each allele of the retinoblastoma gene, in a single susceptible cell before a tumor will develop. This accounts for the later age of onset and the occurrence of single tumors in the familial form of retinoblastoma.[5]

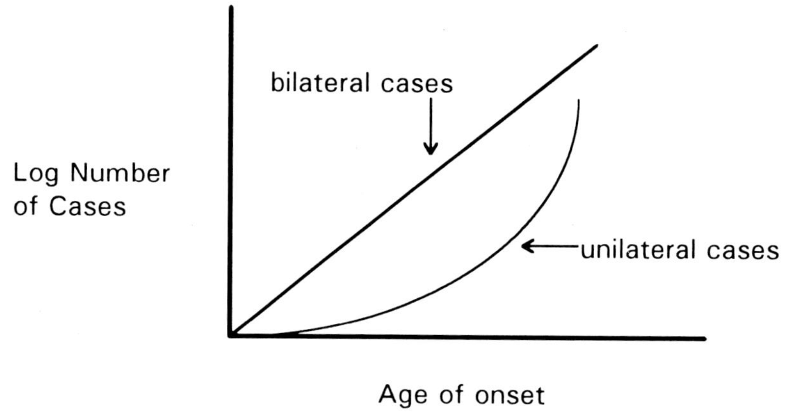

Fig. 2.1. Observations of the age of incidence of bilateral compared to unilateral retinoblastomas led to the "two-hit hypothesis". Bilateral or multifocal tumors occurring in individuals with the familial form of retinoblastoma arose earlier than unilateral or unifocal tumors in the nonfamilial disease. This suggested that individuals with the familial form inherited one of the two requisite mutations and their tumors arose as the result of the second mutation. Individuals without the inherited predisposition to retinoblastomas acquire both mutations postconceptionally and therefore have a later age of onset. (Adapted from Knudson.[1])

Many studies have been undertaken, based on Knudson's two-hit hypothesis, in order to identify tumor suppressor genes. DNA from individuals with a familial predisposition to tumor development has been used in those studies since one mutated allele is already present and can therefore lead to identification of the second mutation. Those studies have led to the discovery of many tumor suppressor genes. Subsequent analyses of DNA from sporadic cases have shown that the same gene is involved in tumor formation in sporadic cases as has been implicated in the familial disease. This has been the case with retinoblastoma, neuroblastoma, Wilms' tumor, the neurofibromatoses type 1 and 2, colorectal cancer, and von Hippel Lindau disease.

B. RETINOBLASTOMA

The retinoblastoma susceptibility gene (*RB1*) was first localized when Knudson and coworkers observed an association between a deletion on the long arm of chromosome 13 and an inherited predisposition to retinoblastoma.[6] The location of the retinoblastoma gene was further refined when two patients were studied who had retinoblastoma, some other mild congenital defects, and an interstitial deletion including band q14 of chromosome 13.[7]

Another study detected deletions of the same region of chromosome 13 using high resolution cytogenetic analyses of tumor cells from a 9-month-old boy with sporadic retinoblastoma.[8] That study detected a homozygous deletion in bands 14q13.2-14.2 in 35% of the tumor cells examined and a heterozygous deletion in an additional 17%. That apparent heterogeneity in the mutations observed in cells of a single sporadic tumor was also reported by Tien and coworkers.[9] In that report cells from a tumor removed from a 23 month old boy were analyzed and three distinct clones were detected suggesting that the tumor had a multifocal origin. The authors caution that such heterogeneity in a single tumor indicates that the disease is familial despite the apparent formation of a single tumor mass. Future diagnoses should be guided by this observation so that follow-up treatment appropriate for the familial disease can be rendered.

In order to test the applicability of the two-hit hypothesis to retinoblastoma, probes had to be isolated which both localize to chromosome 13 and detect restriction fragment length polymorphisms (RFLPs). A collection of probes meeting these criteria was isolated for the study of retinoblastoma.[10] Analyses for loss of heterozygosity were performed using probes from the long arm of chromosome 13.[11] Probes which detect RFLPs on Southern blots revealed a reduction to homozygosity in tumors compared to nontumor tissue from the same patient, thus supporting the contention that biallelic inactivation of the retinoblastoma gene was required for tumor formation (see Fig. 2.2). A definitive test of whether the product of any gene is a tumor suppressor involves the reconstitution of gene function in a malignant cell line which does not express that particular gene. A series of experiments was performed in which a recombinant expression vector carrying the retinoblastoma gene was introduced into neoplastic cells.[12] Cultured osteosarcoma cells (Saos-2) not normally expressing *RB1* displayed morphological changes and reduced tumorigenicity when this recombinant construct was introduced. Similar cells that normally express *RB1* (U-2OS) did not show any changes after introduction of a wild type *RB1* gene.

Historically individuals with retinoblastoma went undiagnosed until the tumor had grown to a size that was visible. By that time the development of metastatic tumors insured an early death for the patient. Once the disease was better understood and diagnostic techniques were developed which permitted early detection, enucleation of the affected eye precluded metastatic disease. Individuals who were treated in that way survived childhood but developed osteosarcomas, most often in their second decade of life. The involvement of the *RB1* gene in osteosarcomas was elucidated by the identification of *RB1* mutations in osteosarcomas.[13,14]

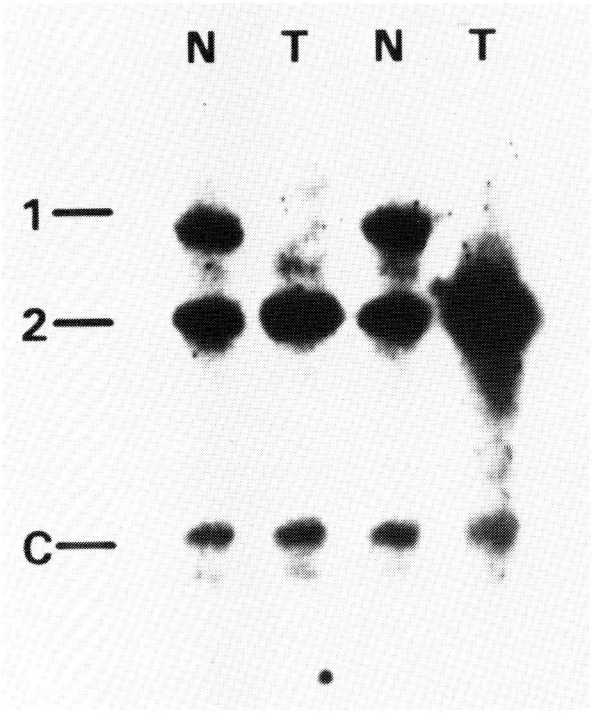

Fig. 2.2. Analysis of tumor suppressor genes by Southern blotting using radiolabeled probes which detect restriction fragment length polymorphisms (RFLPs) demonstrates their loss in tumors. Shown is a Southern blot comparing DNA isolated from normal tissue (N) with that of tumor (T) from the same individuals. The polymorphic alleles detected by this probe are labeled 1, for the larger one, and 2, for the smaller one. A constant band is also detected (C). The two alleles detected in normal tissue demonstrated the constitutional heterozygosity in the individuals analyzed that is lost in their respective tumor. This loss of heterozygosity is a hallmark of tumor suppressor genes.

1. *RB1* Mutations

Detailed elucidation of the function of the *RB1* gene was possible following its isolation by two independent groups.[15,16] The *RB1* gene is composed of 27 exons carried on more than 200 kilobases (kb) of genomic DNA.[17] A large number of mutations have been identified that lead either to a truncated *RB1* mRNA or to its elimination.[17-27] In every instance it is clear that tumors arise, as predicted by the two-hit hypothesis, as a result of biallelic inactivation of the *RB1* gene.

There have also been some reports of *RB1* inactivation by means other than physical mutations, e.g. by hypermethylation.[28] Studies were performed to determine whether genomic imprinting plays a role in the pathogenesis of sporadic retinoblastoma. Genomic imprinting is a process which results in inactivation of some chromosomal regions by hypermethylation. In order to elucidate the potential role of genomic imprinting, the parental origin of new *RB1* mutations was investigated. One study examined the parental origin of the first inactivated *RB1* allele in 13 sporadic osteosarcomas and reported a preference for mutation of the paternally derived *RB1* gene.[29] A similar study was carried out on 15 sporadic retinoblastomas and found no preference for either allele.[30] It is conceivable that, while the *RB1* gene is involved in the pathogenesis of both of these tumor types, genomic imprinting plays a different role in each. A retrospective epidemiological study revealed parental age effect, both maternal and paternal, was observed for sporadic hereditary retinoblastoma.[31] Sporadic hereditary retinoblastoma was defined as the occurrence of bilateral tumors in a patient with no previous family history of the disease or a unilateral tumor in a patient with an affected sibling.

The product of the *RB1* gene is a 105-107kDa nuclear phosphoprotein which plays a role in regulating cellular DNA synthesis.[32-35] Inactivation of the retinoblastoma protein has been observed by mutations which affect phosphorylation, nuclear tethering, and viral oncoprotein association.[36] This last characteristic of the retinoblastoma protein is detailed in chapter 3.

RB1 Mutations in Other Tumor Types

Inactivating mutations of the *RB1* gene have been identified in a variety of tumor types demonstrating that the *RB1* gene product, pRB, plays an important role in regulating cell proliferation in a variety of cell types, not only in retinoblasts.[37] A survey of several tumor types using polyclonal antisera directed against the *RB1* gene product frequently identified aberrant pRB in small cell lung carcinoma and bladder carcinomas with less frequent *RB1* mutations in human mammary tumors and melanomas.[38] Similar findings have been reported by others upon analysis of the *RB1* gene in both primary breast tumors[39] and breast cancer cell lines.[40]

Other studies employed DNA probes to examine the *RB1* gene in small cell lung cancer. Abnormal *RB1* transcripts, either truncated or reduced in abundance, were detected in three of the nine small cell lung carcinoma cell lines studied.[41] Another study detected trace amounts of *RB1* transcripts in 10 of 13 small cell lung carcinoma cell lines studied.[42]

There have been reports of *RB1* mutations in human leukemias. A homozygous deletion of the *RB1* gene was observed in an acute lymphoblastic leukemia cell line.[43] Similar homozygous deletions were observed in two primary leukemias.[44] The first was an acute lymphoid leukemia and the second a chronic myelomonocytic leukemia. Loss of both alleles of the *RB1* gene was also observed in four patients with leukemia and preleukemia.[45] It is interesting that these deletions were found in leukemias of both lymphoid and myeloid origin supporting the important role of the *RB1* gene product in regulating cellular proliferation. This contention was demonstrated by Terada and coworkers when they reported differential regulation of *RB1* expression during cell cycle progression in normal human T cells.[46]

RB1 mutations have also been observed in various sarcomas further demonstrating the role of this tumor suppressor gene in a widening range of tumor types.[47-50] Initially it was believed that pRB was a tissue specific growth suppressor but the repeated demonstration of *RB1* mutations in a variety of tumors has demonstrated its fundamental importance in cell growth regulation.

C. NEUROBLASTOMA

Neuroblastoma in the most common neoplasm of the nervous system occurring in children.[51,52] The role of a tumor suppressor has been implicated in neuroblastoma because of the demonstration of the loss of heterozygosity for probes on chromosome 1 in tumors compared to nontumor from the same patients.[53] The putative tumor suppressor gene responsible for neuroblastoma was further localized to chromosome 1p36 and a preference for loss of the maternal allele was demonstrated by Caron and coworkers.[54] At the same time there is an amplification of the N-*myc* protooncogene located on chromosome 1. The relationship between the loss of a putative tumor suppressor gene and the amplification of a protooncogene is not known. Our understanding of the pathogenesis of neuroblastoma is further complicated by the observation that a component of the nerve growth factor receptor, encoded by the *trk* protooncogene, is highly expressed in some neuroblastomas. High levels of *trk* expression appear to be correlated with a favorable clinical outcome.[55]

D. WILMS' TUMOR

Wilms' tumor is an embryonal tumor of the kidney that can occur in a sporadic or familial form.[56] Wilms' tumor often occurs in a constellation of disorders including Wilms' tumor, aniridia, genitourinary malformations, and mental retardation (WAGR).[57] The gene for Wilms' tumor (*WT1*) lies on the short arm of chromosome 11.[58] The role of the *WT1* gene as a tumor suppressor was supported by the finding of loss of heterozygosity for polymorphic probes identified from the corresponding region of chromosome 11.[56,59,60] Originally it was believed that all of these disorders arose from a defect of a single gene located on chromosome 11p13. However, the involvement of more than one locus was revealed when allelic loss of heterozygosity was detected for probes distal to 11p13 at 11p15.5.[61] A subsequent study looked for loss of heterozygosity for six putative tumor suppressor regions (11p13, 11p15.5, 3p, 5q, 17p, 22q) and found losses only for 11p13 and 11p15.5 in 44 Wilms' tumors analyzed.[62]

There does appear to be a preference for loss of the maternal allele in sporadic cases.[63] DNA from five patients was studied and a demonstrated preferential loss of maternal alleles was observed suggesting that perhaps genomic imprinting or some other epigenetic event may be involved in the sporadic form of this disease.

Surprisingly, the linkage studies of Wilms' tumor failed to show linkage of the familial form of the disease to the previously identified *WT1* locus on chromosome 11.[64,65] The nature of unlinked effector genes which may predispose to mutations at the *WT1* locus has yet to be revealed.

Conclusive evidence that the locus on 11p does indeed encode a tumor suppressor came from a line of experiments in which a normal human chromosome 11 was introduced into a Wilm's tumor cell line and an amelioration of the tumorigenicity was observed.[66] An elimination of tumorigenicity in nude mice was observed as well as changes in growth characteristics of cultured cells which received the normal chromosome 11. The gene for Wilms' tumor susceptibility was isolated by workers in two laboratories nearly simultaneously.[67,68] The protein encoded by the *WT1* gene has four zinc finger domains[69] and functions as a repressor of transcription.[70] One of the genes which is transcriptionally down-regulated by the product of the *WT1* gene is the insulin-like growth factor (IGF-II).[70] The protein encoded by this gene is a potent mitogen active during fetal development and is expressed at elevated levels in Wilms' tumors. Confirmation that this was the gene responsible for Wilms' tumor was obtained when an internal deletion of 25 bases within it was demonstrated in a tumor arising in a patient with a negative family history for this disease.[71]

The *WT1* gene is expressed in a relatively small number of tissues including the developing kidney, genitourinary system, and undifferentiated hematopoietic cells.[72] Expression of the *WT1* gene is regulated at the transcriptional level. Three separate promoters and an enhancer, located 3' of the *WT1* gene have been identified.[72]

While all of the mechanisms which modulate expression of the *WT1* gene or its product have not yet been revealed it is interesting to note that the *WT1* transcript in rats undergoes a relatively rare form of posttranscriptional modification, namely RNA editing. Sharma and coworkers have reported their detection of RNA editing in the *WT1* RNAs of rat.[73] That event results in the conversion of a uracil to a cytosine which in turn leads to the substitution of a proline for a leucine. They report that this change results in a decrease in the ability of the *WT1* encoded protein to repress epidermal growth factor receptor (EGR-1) transcription by approximately 30%. RNA editing is relatively rare in mammals and may represent a mechanism for regulating the activity of tumor suppressor genes which was previously overlooked.

E. NEUROFIBROMATOSIS TYPE 1 (NF1)

Neurofibromatosis type 1 (NF1) or von Recklinghausen's disease is the most common hereditary disease predisposing to neoplasia.[74,75] NF1 is unlike other diseases which predispose to tumor development and involve tumor suppressor genes because of the nontumor clinical features characteristic of NF1 including skeletal abnormalities, learning disabilities, and cafe au lait spots.[76] The gene for NF1 was localized to the long arm of chromosome 17 by genetic linkage analyses.[77,78] Using tightly linked probes which detect restriction fragment length polymorphisms (RFLPs) Skuse and coworkers[79] demonstrated the loss of heterozygosity in malignant NF1 tumors indicating that the NF1 gene was a tumor suppressor. Additional analyses of benign tumors in NF1 failed to detect similar losses.[80]

A major finding which led to the isolation of the NF1 gene was the identification of two NF1 patients who carried translocations, between chromosomes 1 and 17 in one case and chromosomes 22 and 17 in the second, which were responsible for the disease.[81-83] Localization of the translocation breakpoint helped to limit the region which was searched for potential candidates for the NF1 gene. Interestingly a similar translocation was identified in a cell line derived from a malignant Schwannoma, a tumor type common in NF1.[84] That translocation involved the NF1 region of chromosome 17, namely 17q11.2, and the short arm of chromosome 13.

While pathogenetic translocations involving the NF1 gene appear to be rare, several other types of mutations have been identified in NF1. Deletions, point mutations, and insertions appear to be distributed throughout the gene with no particular region appearing to be consistently mutated.[85-97]

The NF1 gene is unusual in a number of ways. In addition to being large, encompassing more than 300kb of the long arm of chromosome 17, it carries within one of its exons three additional genes.[87] One of these genes is a homolog of a mouse gene involved in the pathogenesis of leukemia.[86,98-100] Another, RC1, is a transmembrane protein of unknown function.[87] The third gene encodes the oligodendrocyte-myelin glycoprotein (OMGP), a cell surface protein involved with adhesion.[101] While it is not clear whether any of these genes plays a role in the pathogenesis of NF1 it is intriguing that they are collocated within an NF1 intron and of the two with known function, one is homologous to a murine protooncogene and the other is a cell surface adhesion molecule, a function similar to the tumor suppressor gene, *DCC*, involved in colorectal cancer.[102]

Analysis of the sequence of the NF1 cDNA revealed a region of homology with the mammalian GTPase activating protein (GAP, see Fig. 2.3) and the yeast inhibitor of *ras* proteins (IRA1 and IRA2).[103-105] Introduction of the GAP related domain (GRD) of the NF1 gene into yeast deficient for IRA function resulted in complementation of that deficiency.[105] Mammalian GAP and yeast IRA proteins stimulate the GTPase intrinsic to the product of the *ras* protooncogene. Ras can exist in two forms, either bound to GTP or bound to GDP. In the GTP bound form ras is active in transducing extracellular signals typically arising when ligands bind to receptors on the cell surface. These receptors may be for cell type-specific or more general growth or differentiation factors (see Fig. 2.4). The GDP bound form of ras is inactive in signal transduction. The tumor suppressor function of the NF1 protein, termed neurofibromin, acts by inducing the conversion of the active form of ras, ras-GTP to the inactive form, ras-GDP.

The *NF1* gene encodes several transcripts, two of which differ by the alternate splicing of a single exon termed 23a.[106,107] The 63bp exon 23a is located within the NF1 GRD and results in the insertion of 21 amino acids which change the hydropathic profile of the GRD.[106,108] The two transcripts arising from the alternate splicing of exon 23a, termed type I and type II, are differentially expressed in developing neurons. The type I transcript is predominantly expressed in undifferentiated cells while the type II transcript predominates in differentiated cells.[106] Different levels of expression of the two NF1 transcripts are also observed in different tumor types. For example, type II predominates in brain tumors while type I predominates in normal brain[109] but both tran-

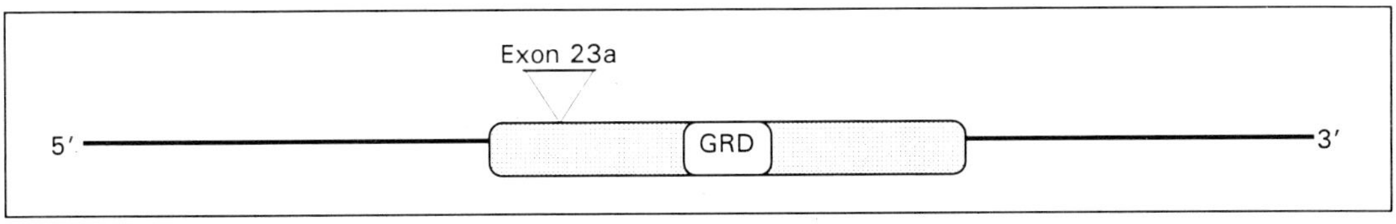

Fig. 2.3. The NF1 mRNA encodes a protein with homology to the GTPase activating proteins (GAPs). The region of homology is indicated by the shaded rectangle labeled GRD (GAP related domain). The inclusion of an alternatively spliced exon, 23a, has an effect on the GAP activity of neurofibromin perhaps by altering the folding of the encoded protein.

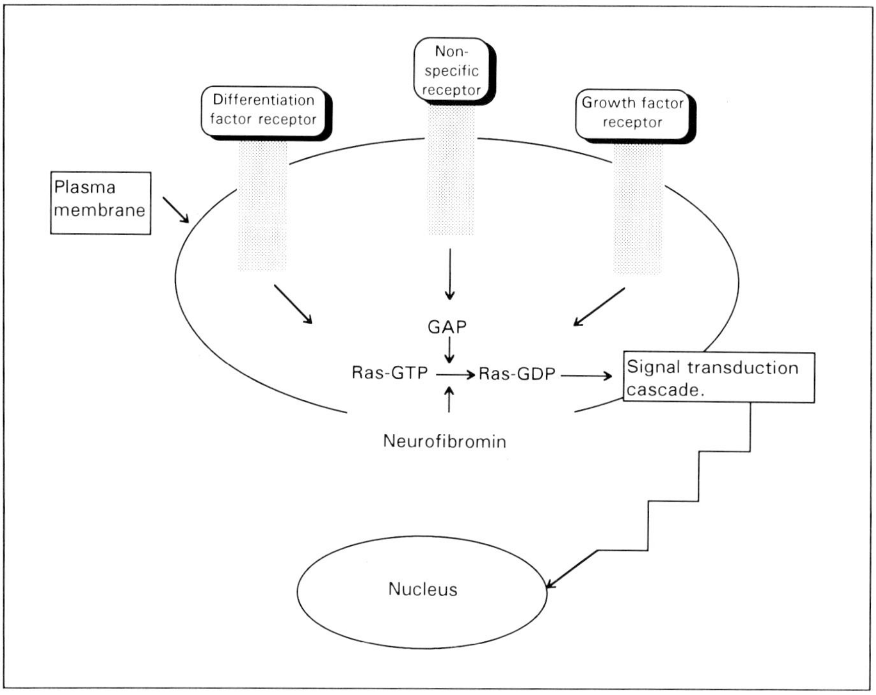

Fig. 2.4. The protein encoded by the NF1 gene, neurofibromin, is a member of the family of GTPase activating proteins (GAPs). It facilitates the conversion of ras-GTP to ras-GDP by the GTPase intrinsic to the ras protein. Since ras-GTP is active in transducing extracellular signals for growth or differentiation, neurofibromin exerts a tumor suppressor effect by converting ras to the inactive form. In certain cells it is likely that neurofibromin is the primary GAP and therefore plays an important role in restraining cell division. Those cells are particularly susceptible to inactivation of neurofibromin and therefore grow unrestrained when it is absent or inactive.

scripts were expressed equally in gastric cancers while the type II transcript predominates in normal stomach mucosa.[110] Peripheral nerve cells and neurofibromas, the most common tumor type occurring in NF1, both express the type II transcript.[111] Interestingly, analysis of the polypeptide encoded by exon 23a reveals homology to nucleoside triphosphatase[111] suggesting that additional function may be intrinsic to the type II transcript.

The protein encoded by the NF1 gene, neurofibromin, was identified using antibodies prepared against synthetic NF1 peptides.[112,113] The protein precipitated by those antibodies has a molecular weight of approximately 220-250kDa, is insoluble in triton X-100 and is expressed primarily in human and bovine brain but not in spleen, thymus, kidney, liver, lung, and heart.[114] Neurofibromin was also detected in several tissues in the rat and mouse.[112] A thorough characterization of the tissue-specific expression of neurofibromin revealed that it is highly expressed in neurons, non-myelinating Schwann cells, oligodendrocytes but not detectable in astrocytes or myelinating Schwann cells.[115] Abnormal regulation of mammalian ras has been reported in malignant Schwannomas resected from NF1 patients[116,117] suggesting that the perturbation of the proposed growth regulatory function of neurofibromin, namely its role in converting ras-GTP to the inactive GDP bound form, may be important in the formation of malignant tumors. Similarly abnormal ras irregularities were not observed in either melanoma or neuroblastoma cell lines carrying inactivated NF1 genes.[118] These contrasting findings in different tumor types suggests that there are multiple regulatory mechanisms affecting ras which may be redundant in some cell lines but not in others.

F. NEUROFIBROMATOSIS TYPE 2 (NF2)

Neurofibromatosis type 2 is clinically and genetically distinct from NF1. It primarily involves tumors of the vestibular branch of the eighth cranial nerve, namely acoustic Schwannomas but affected individuals also may have peripheral neurofibromas, cafe au lait spots, and meningiomas.[119,120] The tissues affected in NF2 are derived from the embryonic neural crest, as are the ones affected in NF1. The occurrence of bilateral tumors is a hallmark of this disease, often resulting in deafness, and along with the dominant mode of inheritance suggested that a tumor suppressor may be responsible for tumor formation due to the multicentricity of the tumors observed.

The gene for NF2 was mapped to the long arm of chromosome 22 by genetic linkage analyses of families with the disease.[119,121] The finding of a constitutional translocation involving chromosome 22q in a patient with NF2 confirmed that localization.[122] Loss of heterozygosity was observed for the region carrying the NF2 gene in a number of meningiomas[123-126] and in NF2 neurofibromas.[126] While there have been several reports that two distinct loci exist on chromosome 22q, one for NF2 and another for meningiomas,[127,128] Ruttledge and coworkers observed the complete inactivation of the NF2 gene in a number of meningiomas.[129]

The NF2 gene was isolated after the identification of mutations, two nonoverlapping deletions and two alterations, in unrelated individuals with NF2.[130,131] The product of the NF2 gene shared homology with a family of proteins which were understood to link integral proteins of the cell membrane with intracellular cytoskeletal components. This proposed structural role of the NF2 gene product, termed Merlin, was distinct from the roles played by several of the other tumor suppressor gene products previously characterized. The potential for NF2 gene mutations in other tumor types was demonstrated when Bianchi and coworkers[132] detected a novel isoform of the NF2 transcript which is differentially expressed in a tissue specific manner and was found mutated in melanoma and breast carcinoma.

G. COLORECTAL CANCER

Among the tumor types for which tumor suppressor gene inactivation has been implicated, colorectal tumors are the most thoroughly studied.[133] One of the reasons for this is that colorectal tumors most often occur in large numbers within a single patient. Among those tumors there are many benign and malignant tumors representing multiple stages along the malignant progression pathway. This situation is fortuitous for researchers because they are afforded the opportunity to study the genetic evolution of malignant tumors in a single patient so that genetic changes relevant to tumorigenesis and progression can be identified against a constant genetic background. The genetic defects leading to malignant colorectal tumors include activation of the *ras* oncogene as well as losses of tumor suppressor activity localized on chromosomes 5q, 17p, and 18q.[134]

Table 2.1. A partial list of known tumor suppressor genes, their subcellular location and their function

Tumor Suppressor	Gene Name	Subcellular Location	Function	References
Retinoblastoma	RB1	Nucleus	Transcription factor binding protein	23
Neurofibromin	NF1	Cytoplasm	GTPase Activating Protein (GAP)	33,34
p53	p53	Nucleus	Transcription factor	25-27
Merlin	NF2	Cytoskeleton	Structural integration of cytoskeleton to plasma membrane	130,131
WT1	WT1	Nucleus	Transcription factor	31,32
DCC	DCC	Plasma membrane	Cell adhesion	35
MCC	MCC	Cytoplasm	Unknown	139
APC	APC	Cytoplasm	β-catenin binding	149
p16	MTS1	Nucleus	Kinase inhibitor	148
VHL	VHL	Unknown	Unknown	146,147

One of the tumor suppressor genes inactivated in colorectal tumors is *p53*,[135] a gene which is involved in many tumor types and plays a pivotal role in regulating cell cycle progression.[136,137] *p53* is typically a secondary or later genetic event in tumor progression but has been implicated as the primary defect in Li-Fraumeni syndrome, a heritable disorder which predisposes to several tumor types.[138] The other two tumor suppressors are DCC (deleted in colon cancer), located on chromosome 18q,[102] and MCC (mutated in colon cancer), located on chromosome 5q.[139] The DCC gene encodes a cell surface protein which shares homology with the neural cell adhesion molecule (NCAM).[102] Despite the fact that the *DCC* gene is expressed in many tissue types *DCC* mutations have been restricted to colorectal tumors. It is interesting then that antisense RNA directed against the *DCC* gene displayed the capacity to transform fibroblasts in culture.[140,141]

H. von HIPPEL LINDAU DISEASE (VHL)

von Hippel Lindau disease is a dominantly inherited syndrome predisposing to tumors of the cerebellum, retina, kidneys, adrenal glands, pancreas, and epididymis.[142] The gene for VHL was mapped to the short arm of chromosome 3 by genetic linkage analysis using DNA from the protooncogene *raf* as a probe.[143] Further localization was provided by analyses using tightly linked polymorphic markers from the same region, 3p25-p26.[144] That study proposed that two distinct tumor suppressor genes existed on the short arm of chromosome 3, one for VHL and another for renal cell carcinoma, a tumor type often associated with VHL. This proposal is not dissimilar to the situation with NF2 in which two tumor suppressor genes may exist, one for the disease and another for a tumor type common to the disease.

A direct demonstration of the presence of the tumor suppressor capacity of genes on chromosome 3p was provided using interspecific microcell hybrids.[145] Microcells were prepared containing fragments of the short arm of chromosome 3 and were fused to a malignant murine cell line and the resultant hybrids were injected into nude mice. Hybrids carrying fragments of chromosome 3p did not produce tumors while the parent cells and hybrids with other chromosomal fragments did.

The VHL gene was ultimately identified using positional cloning techniques.[146,147] Sequence analysis of the transcripts encoded by this gene revealed little homology to anything previously identified except for an acid repeat domain similar to that found in the procyclic surface membrane glycoprotein of *Trypanosoma brucei*.

Confirmation of this gene as the VHL gene was provided by the identification of mutations in several VHL families and in five sporadic renal cell carcinomas.[147] The actual function of the non-mutated VHL gene has yet to be elucidated.

I. SUMMARY

Hereditary diseases which predispose to neoplasia have proven to be excellent systems for the identification of tumor suppressors. The primary mutation is already present at birth (due to the heritable predisposition) so that the mutations directly responsible for tumorigenesis are more easily identified. In reviewing what is known about the various tumor suppressors and their role in the diseases discussed in this chapter several observations become clear. First, tumor suppressor genes are responsible for tumors from disparate tissues. Second, the products of tumor suppressor genes function at many levels within the cell. For example they may be transcriptional regulators (such as pRB or p107), they may play a role in cytoplasmic signal transduction (such as neurofibromin), or they may be cell surface adhesion molecules (such as DCC). Third, some tumor suppressors are involved only in the development of tumors (pRB, p107, WT1) while others play some role in other phenotypic characteristics of disease (neurofibromin). In every case the tumor suppressor protein plays a role in the normal regulation of cellular proliferation or differentiation. Tumors arise when that role is perturbed and cell growth becomes unrestrained.

REFERENCES

1. Knudson AG. Mutation and Cancer: Statistical study of retinoblastoma. Proc Natl Acad Sci USA 1971; 68:820-823.
2. Knudson AG, Strong LC. Mutation and cancer: Neuroblastoma and pheochromocytoma. Am. J Hum Genet 1972; 24:514-532.
3. Knudson AG, Strong LC. Mutation and cancer: A model for Wilms' tumor of the kidney. J Natl Cancer Inst 1972; 48:313-324.
4. Comings DE. A general theory of carcinogenesis. Proc Natl Acad Sci USA 1973; 70:3324-3327.
5. Knudson AG. Model hereditary cancers of man. Prog Nucl Acids Res 1983; 29:17-25.
6. Knudson AG, Meadows AT, Nichols WW et al. Chromosomal deletion and retinoblastoma. New Eng J Med 1976; 295:1120-1123.
7. Yunis JJ, Ramsay N. Retinoblastoma and subband deletion of chromosome 13. Am J Dis Child 1978; 132:161-163.
8. Lemieux N, Milot J, Barsoum-Homsy M et al. First cytogenetic evidence of homozygosity for the retinoblastoma deletion in chromosome 13. Cancer Genet Cytogenet 1989; 43:73-78.
9. Tien H-F, Chuang S-M, Chen M-S et al. Cytogenetic evidence of multifocal origin of a unilateral retinoblastoma. Cancer Genet

904614

marginalia / underlining
highlighting
PON
minor library markings
bookplate
else

—

tcbt
ex lib
~~good~~
good+
very good
fine
new

—

DJ:

new, fine, v.g. good+,
good only, poor

-----

condition

1 2 3 4 11

ship class

1 2 3 4

p m e c b phil hist

R GNT 70

Cytogenet 1989; 42:203-208.
10. Cavenee W, Leach R, Mohandas T et al. Isolation and regional localization of DNA segments revealing polymorphic loci from human chromosome 13. Am J Hum Genet 1984; 36:10-24.
11. Cavenee WK, Dryja TP, Phillips RA et al. Expression of recessive alleles by chromosomal mechanisms in retinoblastoma. Nature 1983; 305:779-784.
12. Huang H-JS, Yee J-K, Shew J-Y et al. Suppression of the neoplastic phenotype by replacement of the RB gene in human cancer cells. Science 1988; 242:1563-1566.
13. Benedict WF, Fung Y-KT, Murphree AL. The gene responsible for the development of retinoblastoma and osteosarcoma. Cancer 1988; 62:1691-1694.
14. Scheffer H, Kruize YCM, Osinga J et al. Complete association of loss of heterozygosity of chromosomes 13 and 17 in osteosarcoma. Cancer Genet Cytogenet 1991; 53:45-55.
15. Friend SH, Bernards R, Rogelj S et al. A human DNA segment with properties of the gene that predisposes to retinoblastoma and osteosarcoma. Nature 1986; 323:643-646.
16. Lee W-H, Bookstein R, Hong R et al. Human retinoblastoma susceptibility gene: Cloning, identification, and sequence. Science 1987; 235:1394-1399.
17. Bookstein R, Lee EY-HP, To H et al. Human retinoblastoma susceptibility gene: Genomic organization and analysis of heterozygous intragenic deletion mutants. Proc Natl Acad Sci USA 1988; 85:2210-2214.
18. Hong FD, Huang H-JS, To H et al. Structure of the human retinoblastoma gene. Proc Natl Acad Sci USA 1989; 86:5502-5506.
19. Friend SH, Horowitz JM, Gerber MR et al. Deletions of a DNA sequence in retinoblastomas and mesenchymal tumors: Organization of the sequence and its encoded protein. Proc Natl Acad Sci USA 1987; 84:9059-9063.
20. Dunn JM, Phillips RA, Becker AJ et al. Identification of germline and somatic mutations affecting the retinoblastoma gene. Science 1988; 241:1797-1800.
21. Gregar V, Passarge E, Horsthemke B. Somatic mosaicism in a patient with bilateral retinoblastoma. Am J Hum Genet 1990; 46:1187-1193.
22. Horowitz JM, Yandell DW, Park S-H et al. Point mutational inactivation of the retinoblastoma antioncogene. Science 1989; 243:937-940.
23. Lee EY-HP, Bookstein R, Young L-J et al. Molecular mechanism of retinoblastoma gene inactivation in retinoblastoma cell line Y79. Proc Natl Acad Sci USA 1988; 85:6017-6021.
24. Dunn JM, Phillips RA, Zhu X et al. Mutations in the *RB1* gene and their effects on transcription. Mol Cell Biol 1989; 9:4596-4604.
25. Murakami Y, Katahira M, Makino R et al. Inactivation of the retinoblastoma gene in human lung carcinoma cell line detected by single-strand conformation polymorphism analysis of the polymerase chain reaction product of cDNA. Oncogene 1991; 6:37-42.

26. Blanquet V, Creau-Goldberg N, de Grouchy J et al. Molecular detection of constitutional deletions in patients with retinoblastoma. Am J Med Genet 1991; 39:355-361.
27. Janson M, Kock R, Nordenskjold M. Constitutional deletions predisposing to retinoblastoma. Hum Genet 1990; 85:21-24.
28. Sakai T, Toguchida J, Ohtani N et al. Allele-specific hypermethylation of the retinoblastoma tumor-suppressor gene. Am J Hum Genet 1991; 48:880-888.
29. Toguchida J, Ishizaki K, Sasaki MS et al. Preferential mutation of paternally derived RB gene as the initial event of sporadic osteosarcoma. Nature 1989; 338:156-158.
30. Dryja TP, Mukai S, Patersen R et al. Parental origin of mutations of the retinoblastoma gene. Nature 1989; 339:556-558.
31. DerKinderen DJ, Koten JW, Tan KEWP et al. Parental age in sporadic hereditary retinoblastoma. Am J Ophthal 1990; 110:605-609.
32. Lee W-H, Shew J-Y, Hong FD et al. The retinoblastoma susceptibility gene encodes a nuclear phosphoprotein associated with DNA binding activity. Nature 1987; 329:642-645.
33. Taya Y, Yasuda H, Kamijo M et al. In vitro phosphorylation of the tumor suppressor gene RB protein by mitosis-specific histone H1 kinase. Biochem Biophys Res Comm 1989; 164:580-586.
34. Howe JA, Mymryk JS, Egan C et al. Retinoblastoma growth suppressor and a 300-kDa protein appear to regulate cellular DNA synthesis. Proc Natl Acad Sci USA 1990; 87:5883-5887.
35. Lin BT-Y, Gruenwald S, Morla AO et al. Retinoblastoma cancer suppressor gene product is a substrate of the cell cycle regulator cdc2 kinase. EMBO J 1991; 10:857-864.
36. Templeton DJ, Park SH, Lanier L et al. Nonfunctional mutants of the retinoblastoma protein are characterized by defects in phosphorylation, viral oncoprotein association, and nuclear tethering. Proc Natl Acad Sci USA 1991; 88:3033-3037.
37. Benedict WF, Xu H-J, Hu S-X et al. Role of the retinoblastoma gene in the initiation and progression of human cancer. J Clin Invest 1990; 85:988-993.
38. Horowitz JM, Park S-H, Bogenmann E et al. Frequent inactivation of the retinoblastoma anti-oncogene is restricted to a subset of human tumor cells. Proc Natl Acad Sci USA 1990; 87:2775-2779.
39. Varley JM, Armour J, Swallow JE et al. The retinoblastoma gene is frequently altered leading to loss of expression in primary breast tumours. Oncogene 1989; 4:725-729.
40. Bookstein R, Lee EY-HP, Peccei A et al. Human retinoblastoma gene: Long-range mapping and analysis of its deletion in a breast cancer cell line. Mol Cell Biol 1989; 9:1628-1634.
41. Yokota J, Akiyama T, Fung Y-KT et al. Altered expression of the retinoblastoma (RB) gene in small-cell carcinoma of the lung. Oncogene 1988; 3:471-475.
42. Hensel CH, Hsieh C-L, Gazdar AF et al. Altered structure and expression of the human retinoblastoma susceptibility gene in small cell lung cancer. Cancer Res 1990; 50:3067-3072.
43. Cheng J, Scully P, Shew J-Y et al. Homozygous deletion of the

retinoblastoma gene in an acute lymphoblastic leukemia (T) cell line. Blood 1990; 75:730-735.
44. Chen Y-C, Chen P-J, Yeh S-H et al. Deletion of the human retinoblastoma gene in primary leukemias. Blood 1990; 76:2060-2064.
45. Hansen MF, Morgan R, Sandberg AA et al. Structural alterations at the putative retinoblastoma locus in some human leukemias and preleukemia. Cancer Genet Cytogenet 1990; 49:15-23.
46. Terada N, Lucas JJ, Gelfand EW. Differential regulation of the tumor suppressor molecules, retinoblastoma susceptibility gene product (Rb) and p53, during cell cycle progression of normal human T cells. J Immunol 1991; 147:698-704.
47. Bookstein R, Rio P, Madreperla SA et al. Promoter deletion and loss of retinoblastoma gene expression in human prostate carcinoma. Proc Natl Acad Sci USA 1990; 87:7762-7766.
48. Reissmann PT, Simon MA, Lee W-H et al. Studies of the retinoblastoma gene in human sarcomas. Oncogene 1989; 4:839-843.
49. Weichselbaum RR, Beckett M, Diamond A. Some retinoblastomas, osteosarcomas, and soft tissue sarcomas may share a common etiology. Proc Natl Acad Sci USA 1988; 85:2106-2109.
50. Cance WG, Brennan MF, Dudas ME et al. Altered expression of the retinoblastoma gene product in human sarcomas. N Eng J Med 1990; 323:1457-1462.
51. Kalwinsky DK, Pratt CB. Pediatric malignant solid tumors. Cancer Chemotherapy & Biol Resp Modifiers 1993; 14:628-640.
52. Goodman SN. Neuroblastoma screening data. An epidemiologic analysis. Am J Dis Child 1991; 145:1415-1422.
53. Fong C-T, Dracopoli NC, White PS et al. Loss of heterozygosity for chromosome 1p in human neuroblastoma: correlation with N-*myc* amplification. Proc Natl Acad Sci USA 1989; 86:3753-3757.
54. Caron H, van Sluis P, van Hoeve M et al. Allelic loss of chromosome 1p36 in neuroblastoma is of preferential maternal origin and correlates with N-*myc* amplification. Nature Genet 1993; 4:187-190.
55. Nakagawara A, Arima-Nakagawara M, Scavarda NJ et al. Association between high levels of expression of the *trk* gene and favorable outcome in human neuroblastoma. N Eng J Med 1993; 328:847-854.
56. Koufos A, Hansen MF, Copeland NG et al. Loss of heterozygosity in three embryonal tumours suggests a common pathogenetic mechanism. Nature 1985; 316:330-334.
57. Housman DE, Glaser T, Gerhard DS et al. Mapping of human chromosome 11: Organization of genes within the Wilms' tumor region of the chromosome. Cold Spring Harbor Symposia on Quantitative Biology 1986; 51:837-841.
58. Compton DA, Weil MM, Jones C et al. Long range physical map of the Wilms' tumor-aniridia region on human chromosome 11. Cell 1988; 55:827-836.
59. Orkin SH, Goldman DS, Sallan SE. Development of homozygosity for chromosome 11p markers in Wilms' tumour. Nature 1984; 309:172-174.

60. Koufos A, Hansen MF, Lampkin BC et al. Loss of alleles at loci on human chromosome 11 during genesis of Wilms' tumour. Nature 1984; 309:170-172.
61. Reeve AE, Sih SA, Raizis AM et al. Loss of allelic heterozygosity at a second locus on chromosome 11 in sporadic Wilms' tumor cells. Mol Cell Biol 1989; 9:1799-1803.
62. Mannens M, Devilee P, Bliek J et al. A. Loss of heterozygosity in Wilms' tumors, studied for six putative tumor suppressor regions, is limited to chromosome 11. Cancer Res 1990; 50:3279-3283.
63. Schroeder WT, Chao L-Y, Dao DD et al. Nonrandom loss of maternal chromosome 11 alleles in Wilms' tumors. Am J Hum Genet 1987; 40:413-420.
64. Grundy P, Koufos A, Morgan K et al. Familial predisposition to Wilms' tumour does not map to the short arm of chromosome 11. Nature 1988; 336:374-376.
65. Huff V, Compton DA, Chao L-Y et al. Lack of linkage of familial Wilms' tumour to chromosomal band 11p13. Nature 1988; 336:377-378.
66. Weissman BE, Saxon PJ, Pasquale SR et al. Introduction of a normal human chromosome 11 into a Wilms' tumor cell line controls its tumorigenic expression. Science 1987; 236:175-180.
67. Call KM, Glaser T, Ito C Y et al. Isolation and characterization of a zinc finger polypeptide gene at the human chromosome 11 Wilms' tumor locus. Cell 1990; 60:509-520.
68. Gessler M, Poustka A, Cavenee W et al. Homozygous deletion in Wilms' tumours of a zinc-finger gene identified by chromosome jumping. Nature 1990; 343:774-778.
69. Evans RM, Hollengerg SM. Zinc fingers: gilt by association. Cell 1988; 309:176-178.
70. Drummond IA, Madden SL, Rohwer-Nutter P et al. Repression of the insulin-like growth factor II gene by the tumor suppressor WT1. Science 1992; 257:674-678.
71. Haber DA, Buckler AJ, Glaser T et al. An internal deletion within an 11p13 zinc finger gene contributes to the development of Wilms' tumor. Cell 1990; 61:1257-1269.
72. Fraizer GC, Wu Y-J, Hewitt SM et al. Transcriptional regulation of the human Wilms' tumor gene (*WT1*). J Biol Chem 1994; 269:8892-8900.
73. Sharma PM, Bowman M, Madden SL et al. RNA editing of the Wilms' tumor susceptibility gene, *WT1*. Genes & Devel 1994; 8:720-731.
74. Crowe FW, Schull WJ, Neel JV "A Clinical Pathological, and Genetic Study of Multiple Neurofibromatosis", pp. 142-146, Thomas, Springfield, IL (1956).
75. Riccardi VM, Eichner JE. "Neurofibromatosis: Phenotype, Natural History, and Pathogenesis," pp. 214-226, Johns Hopkins University Press, Baltimore, MD (1986).
76. Cawthon R, White R. Neurofibromatosis 1. Mol Genet Nervous Sys Tumors 1993; 319-327.
77. Barker D, Wright E, Nguyen K et al. Gene for von Recklinghausen

neurofibromatosis is in the pericentromeric region of chromosome 17. Science 1987; 236:1100-1102.

78. Seizinger BR, Rouleau GA, Ozelius LJ et al. Genetic linkage of von Recklinghausen neurofibromatosis to the nerve growth factor receptor gene. Cell 1987; 49:589-594.
79. Skuse GR, Kosciolek BA, Rowley PT. Molecular genetic analysis of tumors in von Recklinghausen neurofibromatosis: Loss of heterozygosity for chromosome 17. Genes Chrom Cancer 1989; 1:36-41.
80. Skuse GR, Kosciolek BA, Rowley PT. The neurofibroma in von Recklinghausen neurofibromatosis has a unicellular origin. Am J Hum Genet 1991; 49:600-607.
81. Schmidt MA, Michels VV, Dewald GW. Cases of neurofibromatosis with rearrangements of chromosome 17 involving band 17q11.2. Am J Med Genet 1987; 28:771-777.
82. Ledbetter DH, Rich DC, O'Connell P et al. Precise localization of NF1 to 17q11.2 by balanced translocation. Am J Hum Genet 1989; 44:20-24.
83. Menon AG, Ledbetter DH, Rich DC et al. Characterization of a translocation within the von Recklinghausen neurofibromatosis region of chromosome 17. Genomics 1989; 5:245-249.
84. Reynolds, JE, Fletcher JA, Lytle CH et al. Molecular characterization of a 17q11.2 translocation in a malignant Schwannoma cell line. Hum Genet 1992; 90:450-456.
85. Wallace MR, Marchuk, DA, Andersen, LB et al. Type 1 neurofibromatosis gene: Identification of a large transcript disrupted in three NF1 patients. Science 1990; 249:181-186.
86. Cawthon RM, O'Connell P, Buchberg AM et al. Identification and characterization of transcripts from the neurofibromatosis 1 region: The sequence and genomic structure of EVI2 and mapping of other transcripts. Genomics 1990; 7:555-565.
87. Viskochil D, Buchberg AM, Xu G et al. Deletions and a translocation interrupt a cloned gene at the neurofibromatosis type 1 locus. Cell 1990; 62:187-192.
88. Upadhyaya M, Cheryson A, Broadhead W et al. A 90 kb DNA deletion associated with neurofibromatosis type 1. J Med Genet 1990; 27:738-741.
89. Stark M, Assum G, Kaufmann D et al. Analysis of segregation and expression of an identified mutation at the neurofibromatosis type 1 locus. Hum Genet 1992; 90:356-359.
90. Lazaro C, Estivill X. Mutation analysis of genetic diseases by asymmetric-PCR SSCP and ethidium bromide staining: application to neurofibromatosis and cystic fibrosis. Mol Cell Probes 1992; 6:357-359.
91. Li Y, Bollag G, Clark R et al. Somatic mutations in the neurofibromatosis 1 gene in human tumors. Cell 1992; 69:275-281.
92. Troilo P, Strong LC, Little JB et al. Spontaneous and induced levels of chromosomal aberration and sister-chromatid exchange in neurofibromatosis: No evidence of chromosomal hypersensitivity. Mut Res 1992; 283:237-242.

93. Kayes LM, Riccardi VM, Burke W et al. Large de novo DNA deletion in a patient with sporadic neurofibromatosis 1, mental retardation, and dysmorphism. J Med Genet 1992; 29:686-690.
94. Upadhyaya M, Shen M, Cherryson A et al. Analysis of mutations at the neurofibromatosis 1 (NF1) locus. Hum Mol Genet 1992; 1:735-740.
95. Zlotogora J. Mutations in von Recklinghausen neurofibromatosis: An hypothesis. Am J Med Genet 1993; 46:182-184.
96. Anglani F, Murgia A, Bedin S et al. A new disease-causing mutation in the GAP-related domain of the NF1 gene. Hum Mol Genet 1993; 2:1057-1059.
97. Kayes LM, Burke W, Riccardi VM et al. Deletions spanning the neurofibromatosis I gene: Identification and phenotype of five patients. Am J Hum Genet 1994; 54:424-436.
98. Buchberg AM, Bedigian HG, Taylor BA et al. Localization of *Evi-2* to chromosome 11: Linkage to other protooncogene and growth factor loci using interspecific backcross mice. Oncogene Res 1988; 2:149-165.
99. Buchberg AM, Bronell E, Nagata S et al. A comprehensive genetic map of murine chromosome 11 reveals extensive linkage conservation between mouse and human. Genet 1989; 122:153-161.
100. O'Connell P, Viskochil D, Buchberg AM et al. The human homolog of murine *Evi-2* lies between two von Recklinghausen neurofibromatosis translocations. Genomics 1990; 7:547-554.
101. Mikol DD, Gulcher JR, Stefansson K. The oligodendrocyte-myelin glycoprotein belongs to a distinct family of proteins and contains the HNK-1 carbohydrate. J Cell Biol 1990; 110:471-480.
102. Fearon ER, Cho KR, Nigro JM et al. Identification of a chromosome 18q gene that is altered in colorectal cancers. Science 1990; 247:49-56.
103. Xu G, O'Connell P, Viskochil D et al. The neurofibromatosis type 1 gene encodes a protein related to GAP. Cell 1990; 62:599-608.
104. Buchberg AM, Cleveland LS, Jenkins NA et al. Sequence homology shared by neurofibromatosis type-1 gene and IRA-1 and IRA-2 negative regulators of RAS cyclic AMP pathway. Nature 1990; 347:291-294.
105. Ballester R, Marchuk D, Boguski M et al. The NF1 locus encodes a protein functionally related to mammalian GAP and yeast IRA proteins. Cell 1990; 63:851-859.
106. Nishi T, Lee PSY, Oka K et al. Differential expression of two types of the neurofibromatosis type 1 (NF1) gene transcripts related to neuronal differentiation. Oncogene 1991; 6:1555-1559.
107. Bernards A, Haase VH, Murthy AE et al. Complete human NF1 cDNA sequence: Two alternatively spliced mRNAs and absence of expression in a neuroblastoma line. DNA Cell Biol 1992; 11:727-734.
108. Nishi T, Saya H. Neurofibromatosis type 1 (NF1) gene: Implication in neuroectodermal differentiation and genesis of brain tumors. Cancer and Metastasis Rev 1991; 10:301-310.
109. Suzuki, Y, Suzuki H, Kayama T et al. Brain tumors predominantly

express the neurofibromatosis type 1 gene transcripts containing the 63 base insert in the region coding for GTPase activating protein-related domain. Biochem Biophys Res Comm 1991; 181:955-961.
110. Uchida T, Matozaki, T, Suzuki T et al. Expression of two types of neurofibromatosis type 1 gene transcripts in gastric cancers and comparison of GAP activities. Biochem Biophys Res Comm 1992; 187:332-339.
111. Teinturier C, Danglot G, Slim R et al. The neurofibromatosis 1 gene transcripts expressed in peripheral nerve and neurofibromas bear the additional exon located in the GAP domain. Biochem Biophys Res Comm 1992; 188:851-857.
112. Gutmann DH, Wood DL, Collins FS. Identification of the neurofibromatosis type 1 gene product. Proc Natl Acad Sci USA 1991; 88:9658-9662.
113. Hattori S, Ohmi N, Maekawa M et al. Antibody against neurofibromatosis type 1 gene product reacts with a triton-insoluble GTPase activating protein toward *ras* p21. Biochem Biophys Res Comm 1991; 177:83-89.
114. Hattori S, Maekawa M, Nakamura S. Identification of neurofibromatosis type I gene product as an insoluble GTPase-activating protein toward *ras* p21. Oncogene 1992; 7:481-485.
115. Daston MM, Scrable H, Nordlund M et al. The protein product of the neurofibromatosis type 1 gene is expressed at highest abundance in neurons, Schwann cells, and oligodendrocytes. Neuron 1992; 8:415-428.
116. DeClue JE, Papageorge AG, Fletcher JA et al. Abnormal regulation of mammalian $p21^{ras}$ contributes to malignant tumor growth in von Recklinghausen (type 1) neurofibromatosis. Cell 1992; 69:265-273.
117. Basu TN, Butmann DH, Fletcher JA et al. Aberrant regulation of *ras* proteins in malignant tumour cells from type 1 neurofibromatosis patients. Nature 1992; 356:713-715.
118. Johnson MR, Look AT, DeClue JE et al. Inactivation of the NF1 gene in human melanoma and neuroblastoma cell lines without impaired regulation of GTP-Ras. Proc Natl Acad Sci USA 1993; 90:5539-5543.
119. Wertelecki W, Rouleau GA, Superneau DW et al. Neurofibromatosis 2: Clinical and DNA linkage studies of a large kindred. N Eng J Med 1988; 319:278-283.
120. Arai E, Ikeuchi T, Karasawa S et al. Constitutional translocation t(4;22) (q12;q12.2) associated with neurofibromatosis type 2. Am J Med Genet 1992; 44:163-167.
121. Rouleau GA, Seizinger BR, Wertelecki W et al. Flanking markers bracket the neurofibromatosis type 2 (NF2) gene on chromosome 22. Am J Hum Genet 1990; 46:323-328.
122. Martuza RL, Eldridge R. Neurofibromatosis 2. N Eng J Med 1988; 318:684-688.
123. Sanson M, Richard S, Delattre O et al. Allelic loss on chromosome 22 correlates with histopathological predictors of recurrence of meningiomas. Int J Cancer 1992; 50:391-394.

124. Fontaine B, Rouleau GA, Seizinger B et al. Equal parental origin of chromosome 22 losses in human sporadic meningioma: No evidence for genomic imprinting. Am J Hum Genet 1990; 47:823-827.
125. Seizinger BR, De La Monte S, Atkins L et al. Molecular genetic approach to human meningioma: Loss of genes on chromosome 22. Proc Natl Acad Sci USA 1987; 84:5419-5423.
126. Seizinger BR, Rouleau G, Ozelius JL et al. Common pathogenetic mechanism for three tumor types in bilateral acoustic neurofibromatosis. Science 1987; 236:317-319.
127. Cogen PH, Daneshvar L, Bowcock AM et al. Loss of heterozygosity for chromosome 22 DNA sequences in human meningioma. Cancer Genet Cytogenet 1991; 53:271-277.
128. Pulst SM, Rouleau GA, Marineau C et al. Familial meningioma is not allelic to neurofibromatosis 2. Neurol 1993; 43:2096-2098.
129. Ruttledge MH, Sarrazin J, Rangaratnam S et al. Evidence for the complete inactivation of the NF2 gene in the majority of sporadic meningiomas. Nat Genet 1994; 6:180-184.
130. Deprez RHL, Groen NA, van Biezen NA et al. A t(4;22) in a meningioma points to the localization of a putative tumor-suppressor gene. Am J Hum Genet 1991; 48:783-790.
131. Trofatter JA, MacCollin MM, Rutter JL et al. A novel moesin-,ezrin-, radixin-like gene is a candidate for the neurofibromatosis 2 tumor suppressor. Cell 1993; 72:791-800.
132. Bianchi AB, Hara T, Ramesh V et al. Mutations in transcript isoforms of the neurofibromatosis 2 gene in multiple human tumour types. Nat Genet 1994; 6:185-192.
133. Goyette MC, Cho K, Fasching CL et al. Progression of colorectal cancer is associated with multiple tumor suppressor gene defects but inhibition of tumorigenicity is accomplished by correction of any single defect by chromosome transfer. Mol Cell Biol 1992; 12:1387-1395.
134. Fearon ER, Vogelstein B. A genetic model for colorectal tumorigenesis. Cell 1990; 61:759-767.
135. Baker SJ, Fearon ER, Nigro JM et al. Chromosome 17 deletions and p53 gene mutations in colorectal carcinomas. Science 1989; 244:217-221.
136. Levine AJ, Momand J, Finlay CA. The p53 tumour suppressor gene. Nature 1991; 351:453-456.
137. Oren M. p53: the ultimate tumor suppressor gene? FASEB J 1992; 6:3169-3176.
138. Srivastava S, Zou Z, Pirolloa K et al. Germ-line transmission of a mutated p53 gene in a cancer-prone family with Li-Fraumeni syndrome. Nature 1990; 348:747-749.
139. Kinzler KW, Nilbert MC, Vogelstein B et al. Identification of a gene located at chromosome 5q21 that is mutated in colorectal cancers. Science 1991; 251:1366-1370.
140. Lawlor KG, Telang NT, Osborne MP et al. Antisense RNA to the putative tumor suppressor gene "deleted in colorectal cancer" transforms fibroblasts. Ann NY Acad Sci 1992; 600:283-285.
141. Narayanan R, Lawlor KG, Schaapveld RQJ et al. Antisense RNA

to the putative tumor-suppressor gene DCC transforms Rat-1 fibroblasts. Oncogene 1992; 7:553-561.
142. Martz CH. von Hippel-Lindau disease: A genetic condition predisposing to tumor formation. Onc Nursing Forum 1991; 18:545-551.
143. Seizinger BR, Rouleau GA, Ozelius LJ et al. von Hippel-Lindau disease maps to the region of chromosome 3 associated with renal cell carcinoma. Nature 1988; 332:268-269.
144. Seizinger BR, Smith DI, Filling-Katz MR et al. Genetic flanking markers refine diagnostic criteria and provide insights into the genetics of von Hippel Lindau disease. Proc Natl Acad Sci USA 1991; 88:2864-2868.
145. Killary AM, Wolf ME, Giambernardi TA et al. Definition of a tumor suppressor locus within human chromosome 3p21-p22. Proc Natl Acad Sci USA 1992; 89:10877-10881.
146. Latif F, Duh F-M, Gnarra J et al. von Hippel-Lindau syndrome: cloning and identification of the plasma membrane Ca^{++} -transporting ATPase isoform 2 gene that resides in the von Hippel-Lindau gene region. Cancer Res 1993; 53:861-867.
147. Latif F, Tory K, Gnarra J et al. Identification of the von Hippel-Lindau disease tumor suppressor gene. Science 1993; 260:1317-1319.
148. Kamb A, Gruis NA, Weaver-Feldhaus J et al. A cell cycle regulator potentially involved in genesis of many tumor types. Science 1994; 264:436-440.
149. Rubinfeld B, Souza B, Albert I et al. Association of the APC gene product with β-catenin. Science 1993; 262:1731-1734.

CHAPTER 3

Tumor Suppressor Binding to Viral Oncoproteins

A. Background and History of DNA Tumor Viruses

1. Classification

Since the original demonstration of sarcoma induction by a virus of chickens,[1] two distinct classes of cancer causing viruses have been identified in a range of vertebrate species. Definition of these classes was made on the basis of the nucleic acid content of the virus genome (see Table 3.1). For the class of oncogenic or tumor-inducing DNA viruses, the more prominent members include papovaviruses, adenoviruses, and herpesviruses. Retroviruses are the only members of the RNA class of viruses with a demonstrated ability to cause cancer. The kinds of tumors caused by these oncogenic viruses encompass a wide spectrum. For example, polyomavirus causes mainly salivary gland tumors but is also capable of causing a wide range of other malignancies. The retroviruses cause mainly leukemias and sarcomas but can also cause mammary and other tumors. In the field of DNA tumor virus research, considerable effort continues to be expended on studying the structure and function of the oncoproteins belonging to the papovavirus, i.e., polyoma, SV40, papilloma, and adenovirus groups. As such, they serve as outstanding models for studying the mechanisms of virally induced carcinogenesis.

2. Polyomavirus

While studying spontaneously developing mouse leukemias, Gross[2] noted that while most mice injected with filtrates developed

Table 3.1.

Virus Family	Nucleic Acid Content	Oncogenic Members
Adenoviruses	DNA	Numerous
Papovaviruses	DNA	Polyomavirus, SV40, SV40-like human viruses, papillomaviruses
Herpesviruses	DNA	Marek disease virus, Lucke's virus, Epstein-Barr virus, cytomegalovirus*, primate herpesvirus
Poxviruses	DNA	Shope fibroma
Parvoviruses	DNA	None

*Transformation in vitro after ultraviolet irradiation
(Adapted from reference 10)

leukemias, some developed parotid gland tumors. Although this experiment was designed to test the cell-free transmission of these tumors, the data suggested the intriguing possibility that filtrates from leukemic mice contained two distinct viruses each responsible for one of the observed cancer types. Subsequently, Gross obtained evidence for the presence of two virus types by high speed centrifugation of leukemic cell extracts.

Stewart et al.,[3] utilizing this same system, found that the parotid tumor agent was capable of producing neoplasia of different types when injected into newborn mice. In view of this result, the agent was named polyomavirus in reference to its ability to contribute towards the formation of many different types of tumors. The genome of polyomavirus codes for three tumor antigens designated large, middle, and small.[4,5] Functionally, large T appears to enhance the initiation of transformation, but is not required to maintain the transformed state.[6] This latter function is performed by middle T-antigen,[7,8] and may be aided by small T-antigen or a part of the large T-antigen.[9]

3. Simian virus 40

Simian virus 40 (SV40) was discovered by Sweet and Hilleman[10] as an agent that multiplies silently in rhesus monkey kidney cells but was found to produce cytopathic changes in similar cultures from African green monkeys. This virus was later shown to produce sarcomas after injection into newborn hamsters. During this same time period, the Salk and Sabin poliovirus vaccines were developed; both of which depended upon growing the virus in primary rhesus monkey kidney cell cultures. Subsequently, it was reported that both the Salk and Sabin vaccines contained large quantities of this newly discovered simian virus. After ingestion of live polio vaccine, many children continued to excrete SV40 for as long as 5 weeks. The suggestive evidence for an oncogenic role of this virus in humans was provided by epidemiological studies on children whose mothers were administered SV40-contaminated polio vaccine while pregnant.[10] It was determined that the incidence of brain malignancies was higher in these individuals than for a control population which was not similarly exposed to the virus. This observation reinforces the tropism typically displayed for SV40 and SV40-like viruses (see below) for the human brain. Additionally, postnatal administration did not increase the incidence of these malignancies.

SV40 carries the genetic information for two early gene products termed large T-antigen[11] and small t-antigen.[12] The oncogenic property of this virus depends upon the production of these two proteins. T-antigen alone can fully transform cultured cells and induce tumors in rodents when present in sufficiently high concentration. Alone, t-antigen cannot do so, but, can directly or indirectly enhance the transforming activity of limiting quantities of T-antigen.[13] The transforming domain region of T-antigen has been mapped to an N-terminal collinear segment extending from amino acid residues 105 to 114.[14] A map of the functional domain regions of T-antigen is shown in Figure 3.1.

4. SV40-LIKE VIRUSES

Human papovaviruses have been isolated from the brain of a patient with progressive multifocal leukoencephalopathy, termed Wiskott-Aldrich syndrome, and frequently from the urine of immunosuppressed individuals. Designated by the initials of the persons from whom they were isolated, e.g. JC virus;[15] BK virus;[16] they can produce tumors in hamsters and transform hamster cells in vitro. Since 70% of humans have antibodies to the SV40-like viruses, they may be the human equivalent of the widespread SV40 of rhesus cultures and polyoma virus of mice.

5. HUMAN PAPILLOMAVIRUSES

Upwards of 60 different human papillomavirus (HPV) genotypes have been identified so far, and all appear to be associated with distinct anatomic sites.[17] The ones most commonly found in genital and mucosal lesions are HPV type 6 (HPV-6), HPV-11, HPV-16, and HPV-18. These latter two types appear to have a

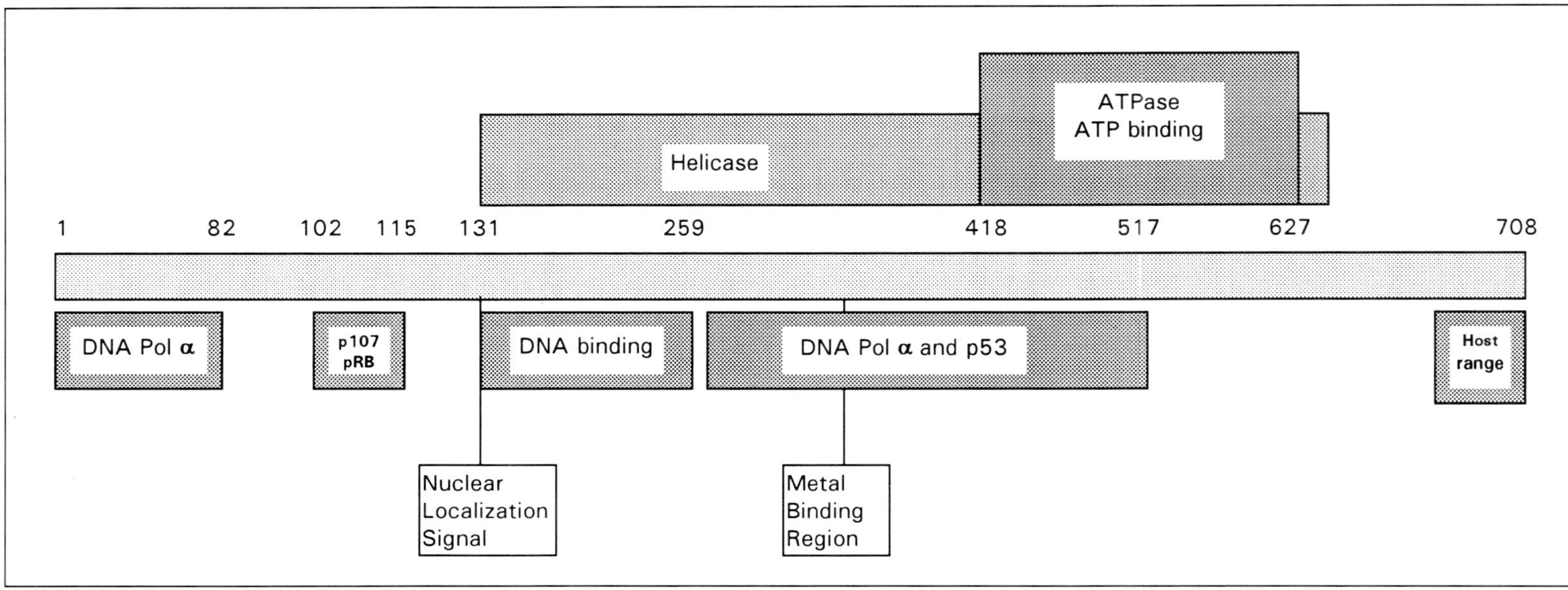

Fig. 3.1. Pictorial representation of SV40 T-antigen functional domains. Shown are the regions involved in binding to DNA polymerase α-primase (DNA Pol α), the tumor suppressor proteins pRB and p53, the cellular p107 protein, metal, ATP, and DNA. Also shown are the regions necessary for ATPase and helicase activities, as well as the nuclear localization signal and the host range domain. Figure was adapted from Fanning.[144]

greater oncogenic potential than the former two as judged by their increased frequency in cervical cancers.[18] Thus, HPV-16 and HPV-18 are considered high-risk types (see Table 3.2). This oncogenicity is also reflected in vitro by their ability to transform established cell lines and immortalize primary rodent cells and human keratinocytes.[19-21] The majority of high-risk associated cancers contain integrated papilloma viral DNA,[22] with the almost invariable retention of the E6 and E7 regions.[23,24] Subsequent studies support the notion that expression of E6 and E7 contributes to malignant cell development. Until recently,[25,26] studies of normal HPV infection, and the role of E6 and E7 in this process, have been difficult to conduct due to the inability to grow HPV in culture. This difficulty has been overcome. Schematic representations of both E6 and E7 are shown in Figure 3.2.

6. Human Adenoviruses

Adenovirus also encodes a set of proteins that disturbs cellular regulatory functions. The virus accomplishes this through the activity of two different genes, designated E1A and E1B. E1A alone can immortalize primary cells in culture,[27] but in cooperation with E1B or an activated *ras* gene can stably transform cultured cells, which in turn can induce tumors in animals.[28,29] Figure 3.3 shows the functional domains of E1A. While human adenoviruses cause a productive infection of human cells and kill the host cells, they can transform nonpermissive hamster or rat fibroblasts. Some can even cause tumors at high frequency (group IV; highly oncogenic) when injected into newborn hamsters while others are weakly oncogenic (group I) or nononcogenic (groups II and III). Viruses from all of these classes can transform hamster cells in vitro. In spite of the wide distribution of adenoviruses, they do not appear to be oncogenic in humans; there is no evidence of human adenovirus DNA being associated with any human neoplasms.[10] This is probably due to the fact that human cells are permissive for the virus; they are killed upon infection rather than transformed.

Table 3.2. Human papillomaviruses associated with gynecological tumors

Type	Cancer Associated	Transforming Ability	Immortilizing Ability
High-risk HPV type 16	Frequently	High	High
High-risk HPV type 18	Frequently	High	High
Low-risk HPV type 6	Rarely	Low	Low
Low-risk HPV type 11	Rarely	Low	Low

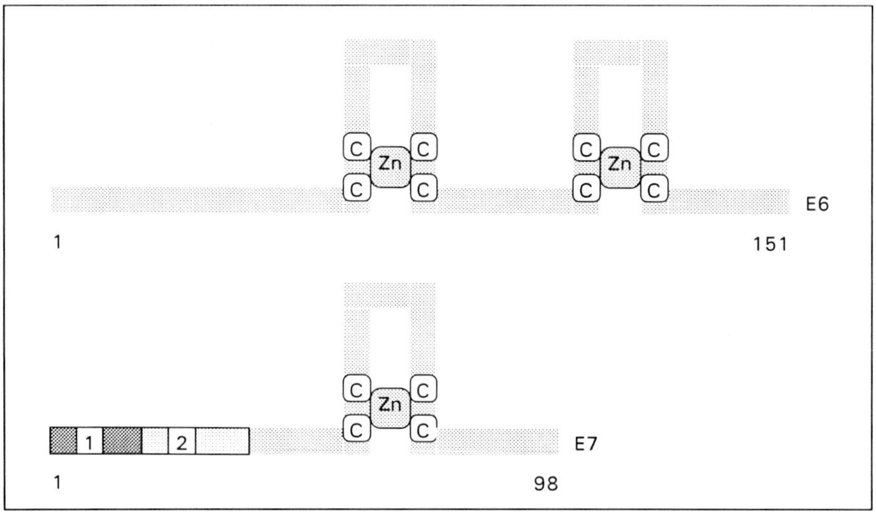

Fig. 3.2. Diagramatic representation of human papillomavirus type 16 E6 and E7 proteins illustrating the Cys-X-X-Cys zinc binding motifs deemed necessary for stability and transforming activity. The highlighted regions in the amino terminus of E7 indicate the regions of sequence similarity between adenovirus conserved regions 1 and 2. Adapted from Vousden.[143]

7. Models for Studying Transformation

The oncogenic potential of these viruses is probably a by-product of the growth mechanisms used by the virus; there is no definitive evidence to support the notion that malignant transformation of the host cell is part of the viral life cycle. Members of these virus classes that induce eukaryotic cells to enter S-phase may consequently induce malignancies; they could override normal cellular controls and maintain cells in a continuously proliferating state. As such, the DNA-containing tumor viruses, particularly those of the papova and adenovirus groups, serve as outstanding models for studying the mechanisms of viral carcinogenesis. As an added bonus, studying the biological activity of their virally coded tumor antigens has provided a wealth of information about the cellular mechanisms involved in normal cell growth and the malfunction of these mechanisms during neoplasia.

B. Viral Oncoprotein Interactions with Growth Suppressor Proteins

The emerging, albeit continuously evolving, picture is that complex formation between the products of viral oncogenes and cellular growth suppressor proteins is associated with an increased propensity towards neoplastic transformation. By studying the molecular biology of such interactions, researchers have acquired significant knowledge with regard to the function of viral oncoproteins during transformation and the role of cellular growth suppressor proteins in cell cycle regulation.

1. T-ANTIGEN BINDING TO THE P53 CELLULAR PROTEIN

While first identified as a cellular protein present in a complex with SV40 T-antigen, p53 has the ability to function as both a promoter and a suppressor of cell growth.[30] While the large T-antigen of polyomavirus does not bind p53, T-antigen of the related lymphotropic papovavirus does.[31] A nuclear phosphoprotein, p53 is not very stable in untransformed cells, having a half-life of around 15-20 minutes.

SV40 T-antigen is also a nuclear phosphoprotein. The nuclear localization of both p53 and T-antigen is controlled in a cell cycle dependent fashion and is apparently essential for their respective activities.[32] p53 molecules from SV40-transformed murine and primate cells are phosphorylated on specific serine residues as is the p53 protein of cells which do not express T-antigen.[33] It has been suggested that SV40 transformation of murine cells results in a specific increase of both T-antigen and p53 phosphorylation correlating with their ability to complex with one another. Indeed, the observation that there is an increase in specific phosphorylation sites in both complexed

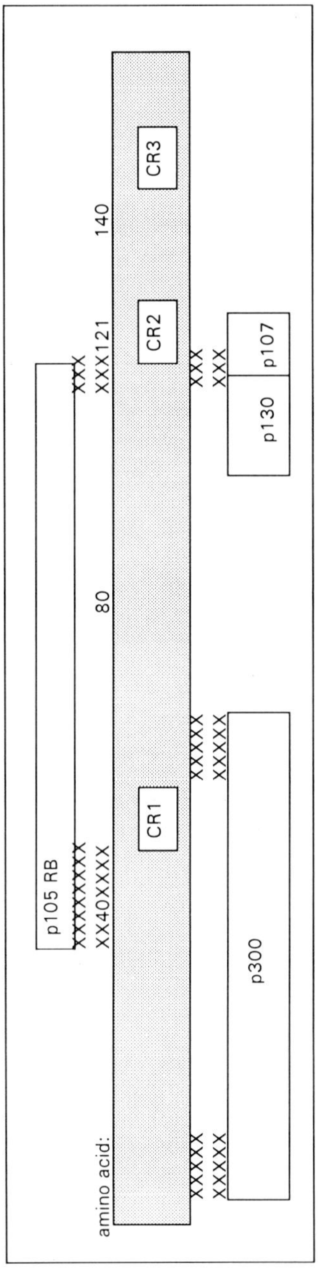

Fig. 3.3. Representation of the different functional domains of adenovirus E1A. Extensive quantitative analysis using E1A deletion mutants have enabled identification of unique binding patterns, contact regions denoted by a series of X's, for pRB, p107, p130, and p300. p300 appears to require sequences in the amino terminus of E1A in addition to sequences in conserved region 1 (CR1). pRB, p107, and p130 appear to react primarily with conserved region 2 (CR2), however, sequences in CR1 and the amino terminus may be capable of weak interactions and appear to function cooperatively with CR2 to bind these proteins. Adapted from Boulanger and Blair.[52]

T-antigen and p53 indicates either an increased affinity of p53 for highly phosphorylated T-antigen or this enhanced phosphorylation may be a consequence of complex formation.[34]

The kinetics and location of complex formation between p53 and T-antigen has been shown to occur very rapidly in the nucleus.[35] Using SV40-transformed mouse and rat cells, it was shown that both proteins were transported to the nucleus, in uncomplexed form, with newly made p53 transported more rapidly than T-antigen. The $t_{1/2}$ for association between newly made p53 with T-antigen was shown to be 5 to 15 minutes, while complex formation between newly made T-antigen with p53 varied from 5 to 30 minutes depending upon the ratio of T-antigen to p53. Residues 165-199, and 232-285 of murine p53 appear to be required for efficient T-antigen binding.[36]

In order to map the p53 binding site on T-antigen, p53 was complexed in vitro to labeled proteolytic fragments of T-antigen.[37] A 46 kDa fragment, spanning residues 131-517, bound and therefore likely contains the p53 binding sites. When combined with additional data using deletion mutants of T-antigen, the p53 binding domain of T-antigen was found to be contained in residues 272-517 (see Fig. 3.1). This region is also necessary for the efficient transformation of nonpermissive cells and for the induction of cellular rRNA synthesis. It also contains the binding sites for DNA polymerase-α and ATP suggesting a possible role for T-antigen/p53 complexes in T-antigen-associated functions. In lytically infected monkey cells, comparisons have been made for the ATPase, DNA-binding, and helicase activities of free T-antigen to that complexed with p53. These activities have been shown to be enhanced for T-antigen in complex with p53.[38] The implication of this finding is that complex formation between T-antigen and p53 modulates a number of essential T-antigen associated activities crucial for a productive infection by SV40. Removal of T-antigen amino acid residues 451-532 appears to alter the capacity of newly synthesized but not older T-antigen and p53 to form complexes.[39]

In further support of this notion, a change from Pro to Leu at position 584 in the T-antigen hydrophobic region resulted in loss of p53 binding concomitant with changes in ATPase activity, oligomerization, stability, and in vivo phosphorylation and can be correlated with defective replication and restricted transforming functions of T-antigen.[39] That so many biochemical changes are associated with a single amino acid substitution in the hydrophobic region of T-antigen is consistent with its importance in regulating higher-order structural and functional relationships in this protein. Additionally, it has been shown that wild-type murine p53 can block the binding of DNA polymerase-α to wild-type T-antigen and that purified p53 dramatically inhibits the ability of T-antigen to mediate the replication of a plasmid bearing the viral origin while high concentrations of mutant transforming p53 failed to block this replication.[40] Thus, it would appear that murine p53 interferes with interactions between T-antigen molecules that are required for DNA synthesis. Taken together, these results suggest that p53 may play a role in restricting the host range of SV40 virus. Mechanistically, this restriction may involve wild-type p53 blocking the binding of a T-antigen-like host initiator protein to DNA polymerase-α to control host cell DNA replication. Absence of p53 expression or mutation of p53 to a non-T/non-host protein binding form would release the cell from this control.

It has been a widely held view that the reason for increased stability of the p53 protein observed in SV40-transformed cells is that complex formation with T-antigen stabilizes the cellular protein. To the contrary, experiments performed using temperature-sensitive mutants of T-antigen have shown that neither complex formation of p53 with T-antigen or T-antigen expression is sufficient for a significant metabolic stabilization of p53.[41] This suggests that metabolic stabilization of p53 during cellular transformation with SV40 is mediated by a cellular process and probably is the consequence of the T-antigen-induced transformed phenotype. Indeed, the phosphorylation of p53 may induce a change in its conformation. This change may lead to stabilization of the anti-proliferative conformation and/or stabilization of turnover rate thus permitting the anti-proliferative form to function. p53 stabilization may also be dependent on its ability to self aggregate. The ability of T-antigen to form stable complexes with p53 does appear, however, to be necessary for SV40 to extend the life span and partially transform human cells in culture.[42] One should not assume from this, however, that complex formation between these two is sufficient for the oncogenic potential of papovavirus T-antigens. Indeed, other cellular protein binding to T-antigen also plays a role.[43,44]

Using a conformation-dependent p53 antibody which can usually distinguish between p53 with anti-proliferative activity and the mutated form with oncogenic activity, it was found that SV40-mediated transformation of murine cells is not dependent upon activation of p53 to the oncogenic form. Instead, transformation is most likely facilitated because the anti-proliferative activity of p53 is blocked by T-antigen binding.[45] The major obstacle in defining this activity is that our present knowledge about p53 is limited in the biochemical characterization of this protein, although new findings continue to be reported in this regard (see chapter 4). When there is a convincing assignment of a proper biochemical activity to p53, interpretation of the different types of mutations and the corresponding phenotypic response of the cell will be greatly facilitated. One step towards the goal of elucidating this anti-proliferative activity of p53 involves the study of SV40 transformants of the human osteosarcoma cell line HOS.[46] These cells express both T-antigen and p53 efficiently, but without apparent molecular association with each other. These studies have revealed that p53 is underphosphorylated and shows unusual peptide fingerprints compared to normal p53 which can associate with T-antigen indicating, perhaps, the functional significance of this p53 modification. It is hoped that this system can be exploited to further define the roles of the respective antigens during progressive malignant transformation.

Studies involving T-antigen mutants have also contributed to our understanding of the role of T-antigen/p53 complex formation in neoplasia.[47] DNA constructs which encode T-antigen lacking the first 109, 127, 150, and 176 amino acids, or internal segments between 117 and 250, were tested for the ability to immortalize C57Bl/6 mouse embryo fibroblasts. Those results indicate that neither the first 127 amino acids or amino acids 127-250 of T-antigen are needed for efficient immortalization of these primary cells or their tumorigenicity. Tumorigenicity was found to be separable from the transformed characteristics of growth in reduced serum, anchorage independent division, and overcoming inhibition of cell division due to high cell density. Thus, neither the portion of T-antigen (131-159) that appears sufficient for SV40 *ori* binding nor the portion (147-214) involved in DNA binding were found to be essential for either of these transformed cell characteristics. Similarly, phosphorylation of T-antigen at residues Ser-106, 112, 123, and Thr-124 and the nuclear localization signal comprised of amino acids 126-132 did not influence efficient immortalization or tumorigenesis. A requirement for residues 251-626 was found for immortalization, suggesting a correlation between p53 binding and the immortalizing capability of T-antigen and that T-antigen binding to p53 may be required for multiple transformed cell growth properties. A more recent study has implicated the amino terminal portion of T-antigen as being required to overcome p53-mediated growth arrest of cells.[48] Data are provided which indicate that two functions within this region are essential to induce cell growth—the ability to bind to pRB and the presence of a domain in the first exon which appears to interact with p300. The authors suggest that T-antigen association with pRB, p300, or both can communicate with the functions of wild-type p53 in a cell.

While initial studies focused only on the effect that T-antigen/p53 complex formation had on the ability of T-antigen to modulate transcription, p53 also appears to have a similar ability. Wild-type p53 can bind specifically to DNA sequences within SV40 as well as several cellular genes. The proposed function of this binding is to modulate transcription of the target genes. It has been reported that such site-specific binding of p53 to cellular DNA can be inhibited by complex formation with T-antigen.[49] This implies that the oncogenic potential of T-antigen involves, at least in part, the ability of complex formation to block binding of wild type p53 to cellular DNA. Another known activity of p53 is transcriptional activation. T-antigen binding to p53 has been shown to inhibit transactivation by wild-type p53, suggesting that transformation by T-antigen may be linked to its ability to inhibit such p53-mediated transcriptional transactivation,[50] some of which may not be dependent upon a stable association between T-antigen and p53.[51]

2. Adenovirus E1B binding to p53

E1B encodes a 55kDa and a 19kDa protein product.[52] The 55kDa protein from adenovirus type 5 (Ad5) can be found complexed with p53. No consensus site for p53 binding to the DNA tumor virus proteins T-antigen, HPV E6, and Ad5 55kDa E1B has been identified. In fact, p53 binding studies have revealed that each of these viral proteins interacts with this cellular protein through a different region of p53.[53] Such binding may not be conserved among adenoviruses. Case in point, the 54kDa E1B product of Ad12 shows no evidence of p53 association.[54] However, the large E1B products from both serotypes appear to influence the function of p53 differently. E1B from the nononcogenic Ad5 serotype forms a complex with p53 and sequesters it in the cytoplasm. The presence of homologous E1B from the oncogenic Ad12 serotype appears to metabolically stabilize p53 in these cells, resulting in a high level of this protein in the nucleus, and contributing to the oncogenicity of these cells.[55] While such properties are generally associated with mutant forms of p53, the stable form of p53 in the Ad12 transformed cells is wild-type, and its stability is a direct consequence of Ad12 E1B expression.

Mutational analysis of the 55kDa transformation deficient E1B suggests that blocking the transcription stimulating activity of p53 is an essential step in the mechanism through which E1B contributes to the transforming activity of adenovirus.[56] Biochemical and genetic approaches have shown that the 55kDa E1B of Ad5 is phosphorylated at serine and threonine residues near the carboxy terminus within sequences characteristic of substrates of casein kinase II.[57] Mutating those residues decreased viral replication and the efficiency with which it transforms primary baby rat kidney cells, while interaction with p53 was little affected and p53 transactivation activity partially inhibited. These results suggest that phosphorylation of these residues either regulates the inhibition of p53 or regulates some other E1B function required for cell transformation. Interestingly, the Ad12 55kDa E1B, which shows no evidence of complexing with p53, was still capable of blocking p53 transcriptional activity.[54] Taken together, these results suggest that, while blocking p53 mediated transcriptional activity is crucial for the transforming ability of 55kDa E1B, actual complex formation is not necessarily required.

The 19kDa E1B product has been implicated with the suppression of the apoptosis-inducing ability of p53.[58] Thus, both products of the adenovirus E1B gene target pathways responsive to p53, albeit by different mechanisms, as there is no known structural homology between the two proteins.

3. Human papillomavirus E6 binding to p53

E6 is the second major oncoprotein encoded by HPV. Experiments have shown that E6 from the high-risk HPV-16 and HPV-18 can associate with wild-type p53.[59] In vitro, such an interaction between E6 from the oncogenic HPVs that bind p53 has been shown to stimulate the degradation of this cellular protein.[60] This degradation is ATP-dependent and involves the ubiquitin-dependent protease system. The interaction between E6 and p53 does not appear to be a direct one; the E6 oncoprotein binds to a cellular protein of 100kDa, termed the E6-associated protein (E6-AP), and it is this complex which specifically interacts with p53.[61] Subsequent cloning of E6-AP has enabled functional mapping studies to be carried out thus localizing the regions of this protein that direct HPV E6 binding, association with p53, and ubiquitination of the associated protein[62] (see Fig. 3.4).

Recent in vivo studies have shown that newly synthesized p53 in E6-expressing human epithelial[63] and keratinocytes[64] is less stable than their non-E6 expressing counterparts. Recently, a novel species of ubiquitin-carrier protein has been found, designated E2-F1, it

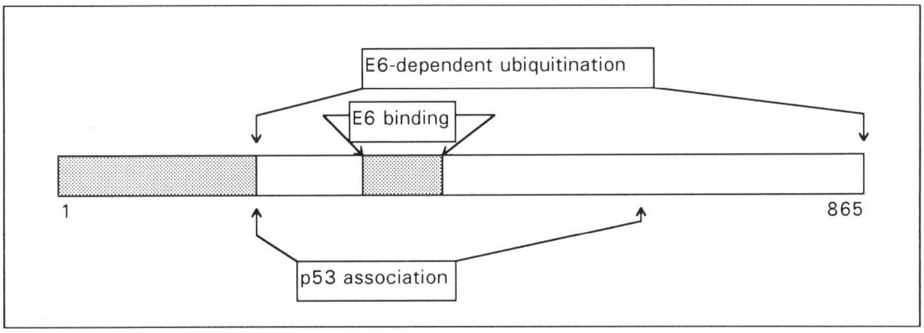

Fig. 3.4. Schematic diagram illustrating domain regions of E6-AP. Shown are the regions responsible for human papillomavirus E6 binding, E6-dependent association with p53, and E6-dependent ubiquitination of p53. Adapted from Huibregtse et al.[62]

is involved in the conjugation and degradation of p53.[65,66] The involvement of this protein in vitro appears to reproduce the in vivo process; mutant p53 that cannot bind E6 and is stable in vivo is not recognized by the cell free system. Thus, association of E6 from the oncogenic HPVs with p53 may lead to loss of function of the cellular protein by eliminating the gene product. While E6 from the low-risk HPV-6 and HPV-11 do appear to form a complex with p53, albeit at a lower affinity, such a complex does not target p53 for degradation.[67] Work from this same laboratory recently showed that E6 from types 16 and 6 can modulate the transcriptional regulatory function of p53 in several cell types.[68] Using a series of mutants, the requirement for this association was demonstrated, and, while mutants which associated but did not target p53 for degradation showed some of this ability to abrogate p53 mediated transcriptional regulation in vitro, the activity was augmented by E6 which did target p53 for degradation. This suggests that some modulation of p53 function is necessary in the normal life cycle of the virus, and that there is a correlation between the efficiency of this activity and the oncogenic potential of the virus.

What could be some of the functional consequences of this interaction promoting p53 degradation? For one, the normal response of cervical epithelial cells to DNA damage appears to be undermined by interactions between E6 produced by the high-risk HPVs and p53.[69] Important cellular responses to such damage which are mediated by p53 are disrupted, and this may contribute to the subsequent accumulation of genetic changes associated with cervical tumorigenesis. E6 binding to p53 can also abrogate p53-mediated repression of transcription.[70] Thus, with the importance of p53 in normal cell division control, most likely at the transcriptional level, E6 binding to p53 would facilitate its degradation, thus relieving its negative control on transcription and, by extension, its block to cell cycle progression. While p53 mutations are frequent in many human cancers, analysis of the p53 sequence in HPV positive cancers revealed that such mutations are rare,[71] suggesting that such cancers display an alternative mechanism for inactivating p53 function without mutations within cellular genes. While clonal p53 mutations are almost never detected in the above mentioned cancers, one study has provided evidence that mutations of p53 are sometimes found in metastases from these tumors.[72] Additional analyses revealed that mutations in p53 could be detected at very low levels in the primary cancer. Thus, it is possible that mutant p53 contributes to the further progression of a primary cancer that initially contained only wild-type p53 which was inactivated at the protein level by E6.

4. Adenovirus E1A complex formation with pRB

The E1A oncogene of adenovirus functions in several biochemical capacities. Perhaps the function which is studied most rigorously is its ability to stimulate unrestricted growth in quiescent primary rodent fibroblasts, a property which is dependent to some degree on its ability to interact with several classes of cellular DNA-binding domains and thereby be recruited to diverse promoters.[73] E1A targets a promoter by interacting with the DNA-binding domain of a cellular transcription factor. The activation regions of E1A and the cellular protein then cooperate with one another to stimulate transcription.

The ability of E1A to bind to specific cellular proteins and perhaps modulate the biological activity of these proteins by enhancing their phosphorylation[74] has attracted the most attention of late. This ability to bind to host cellular proteins appears to depend upon the integrity of two independent transforming domain regions.[75] Thus, E1A is capable of inducing dramatic changes in host cell transcription which facilitates induction of cell cycle activity.[76] Along this line, E1A has recently been proposed to mediate its mitogenic effect by activating the expression of genes for cyclin A and E, both of which are known to contribute towards cell cycle progression.[77]

The retinoblastoma susceptibility gene, *RB1*, which when deleted or otherwise inactivated, predisposes to the development of malignant retinoblastoma.[78] Deletion or inactivation of this gene has also been observed in nearly all small cell lung cancers, some breast cancers, some bladder carcinomas, and certain osteogenic sarcoma cell lines. Reintroduction of the wild type *RB1* allele into

RB1$^{-/-}$ cells, in some instances, led to a decrease in cell growth and a reduction in the ability to form tumors in nude mice.[79] Thus, while *RB1* function seems to be required for the prevention of a specialized tumor of the eye, it may also have neoplastic growth suppression properties in a wider variety of cell types. The product of the *RB1* gene, pRB, has been detected in virtually all mammalian cell types examined and appears to be required for normal mouse development.[80-82] This nuclear protein exists as a differentially phosphorylated family of peptides.[83] Given that pRB is present in virtually all normal cell types, it can be argued that pRB functions to suppress growth in a wide variety of cells.

Towards the latter part of the 1980s, investigators observed that the E1A protein of adenovirus could form a specific complex with several host cell proteins[84,85] (see Fig. 3.3). Shortly thereafter, it was reported that one of these proteins, having a molecular weight of 105kDa, was the product of the retinoblastoma susceptibility gene.[86] The identity of this protein was ascertained by several criteria. E1A-associated proteins were used as immunogens. An antibody directed against the 105kDa protein was then used for further studies. This antibody coprecipitated a protein identical in molecular weight and peptide map to the pRB protein recognized by an anti-trp/pRB-fusion protein antibody.[87] This antibody did not recognize any proteins in the 105kDa region from lysates of retinoblastoma cells. Nor could an E1A coprecipitating protein of 105kDa be detected from such cells. Using a series of deletion mutants, the regions of E1A participating in this reaction were mapped.[88] It was reported that pRB required the presence of sequences from two noncontiguous regions of E1A from amino acid residues 30-60 and 121-127; regions which coincide with the transforming function of this viral protein. Using synthetic peptides corresponding to portions of these regions, it was shown that each of these regions can bind independently to pRB.[89] These results suggest that interaction with this cellular protein is fundamental to the transforming activity of E1A, at least as far as initiation of this process is concerned.[90] The interaction between these two proteins was the first demonstration of a physical link between an oncogene and an anti-oncogene or tumor suppressor gene.

5. SV40 T-ANTIGEN BINDING TO PRB

Although T-antigen has been the focus of extensive biological and biochemical characterization for at least the last two decades, its precise role in neoplastic transformation has remained elusive. Insight into how T-antigen functions in this regard was facilitated by the studies demonstrating that complex formation with E1A depended upon the intact nature of a collinear sequence residing in this viral protein.[88] Since an analogous sequence can be identified in SV40 T-antigen[14] (see Fig. 3.5), shortly thereafter, investigators found in the course of immunoprecipitation studies of lysates from monkey cells synthesizing T-antigen that T-antigen and pRB could be coprecipitated in a trimolecular complex with p53[91-92] (see Fig. 3.6).

Substitution and internal deletion mutants of T-antigen that contain structural alterations in the transforming domain region failed to complex pRB. A 14-residue peptide, corresponding to this T-antigen transforming domain region, is able to compete effectively with T-antigen for binding to pRB whereas a peptide with the same sequence, but with a Lys to Glu substitution at position 107, is not.[93] This latter peptide is a replica of the segment in the SV40 T-antigen mutant designated K1, a stable but transformation- and pRB binding-deficient derivative. Taken together, these results are consistent with a model for transformation by SV40 which involves complex formation of the viral T-antigen oncoprotein with pRB and the perturbation of pRB protein and/or T-antigen function.

One caveat of this model when originally proposed was that the T-antigen transformation data were all obtained in mouse and rat cell lines while the T-antigen/pRB complex formation data was from monkey and human cells. When attempts were first made to detect T-antigen/pRB complexes in murine cells, the results were less convincing than those obtained from the monkey and human cell lines. Although mouse cell *RB1* has a high degree of sequence homology with the human gene, and stable murine *RB1* message

Sequence homology among regions of adenovirus E1A, SV40 T-antigen, and HPV E7 essential for pRB binding:

E1A	121AspLeuThrCysHisGluAlaGlyPheProProSerAspAspGluAspGlu137
T	102AsnLeuPheCysSerGluGluMetProSerSerAspAspGluAlaThrAla118
HPV 16	21AspLeuTyrCysTyrGluGlnLeuAsnAspSerSerGluGluGluAspGlu37
HPV 18	24AspLeuLeuCysHisGluGlnLeuSerAspSerGluGluGluAsnAspGlu40

Fig. 3.5 Regions of adenovirus E1A, SV40 T-antigen, and human papillomavirus types 16 and 18 that share amino acid sequence homology essential for complex formation with the tumor suppressor protein pRB. Numbers at the beginning and end of the sequence denote the span of the homology within the entire viral protein. Adapted from Barbosa et al.[113]

could be detected, it was believed that the affinity of the available immunological reagents for the mouse protein was low. By increasing many-fold the amount of immunocomplexed material routinely used for western blotting human and monkey cell samples, an association could then be repeatedly demonstrated using mouse cell extracts.[94] With the availability of additional commercial immunologic reagents, murine pRB complexed with T-antigen can now be more efficiently and reliably detected.

The region of pRB required for T-antigen binding extends from residue 379 to 792,[95] the so called "pocket" binding region. While a family of pRB proteins can be immunoprecipitated from human and monkey cell extracts, only one member can be found coprecipitating with T-antigen. Radiolabel experiments demonstrated

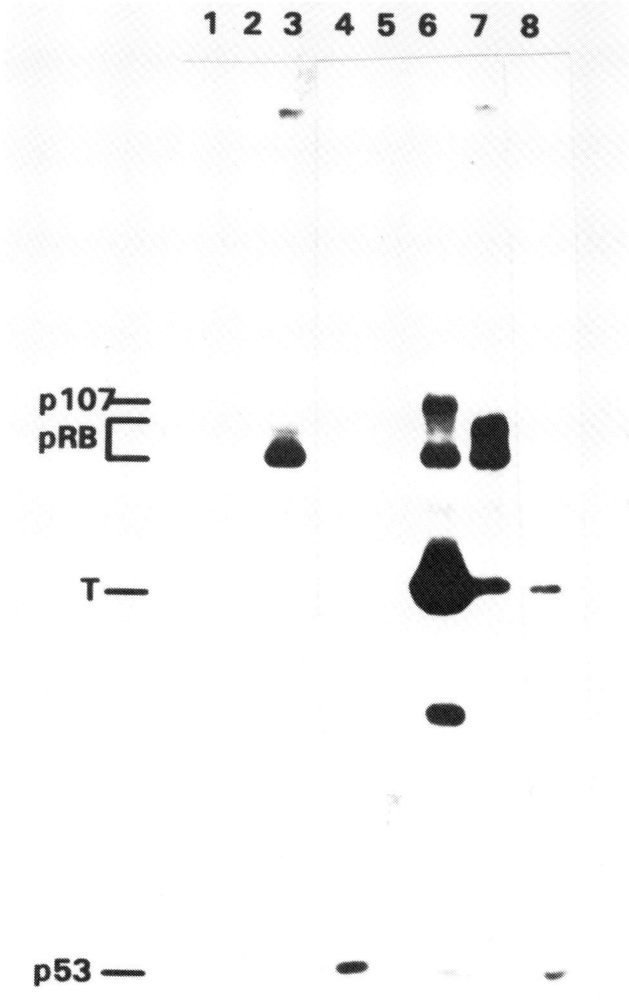

Fig. 3.6. Immunoprecipitation of control and SV40 T-antigen-containing cell extracts. CV-1P monkey kidney cells, a non-T-antigen expressing line used as a control, and D_2C_2 cells, a clonal derivative of CV-1P stably transformed by a plasmid encoding wild-type T-antigen, were radiolableled with ^{35}S-methionine. Extracts were then prepared from these CV-1P (lanes 1-4) and D_2C_2 (lanes 5-8) cells for immunoprecipitation with a null, non-reactive monoclonal antibody (lanes 1 and 5), or monoclonal antibody to T-antigen (lanes 2 and 6), pRB (lanes 3 and 7), or p53 (lanes 4 and 8). Positions of p107, phosphorylated (slower migrating band in the doublet) and underphosphorylated (faster migrating band in the doublet) pRB, as well as p53, are indicated. Radiolabeling and immunoprecipitation was performed as described in Ludlow et al.[83]

that the fastest migrating form which coprecipitates with T-antigen is underphosphorylated with respect to the slower migrating forms of pRB which are overtly phosphorylated.[83] The question then arises as to why the selection for this underphosphorylated form of pRB binding to T-antigen. In keeping with the hypothesis that complex formation between T-antigen and pRB does indeed perturb the function of pRB, one explanation for this apparent selectivity in binding is that the active form of pRB with respect to growth suppression is indeed the underphosphorylated form. By microinjecting purified underphosphorylated pRB, containing the T-antigen binding region, into cells, a formal test of this model was carried out.[96] While microinjection of this protein into early G1 cells was shown to inhibit progression into S, no such inhibition was observed when similar experiments were performed in T-antigen expressing cells. The conclusion here is that T-antigen complexes with the underphosphorylated pRB thus removing the block to exit G1. These data directly support the hypothesis that T-antigen binding has functional consequences for pRB.

The most plausible explanation for why pRB exists in both overtly phosphorylated and underphosphorylated forms is that the function of this protein is modulated by such a posttranslational modification. To address this issue, the phosphorylation state of pRB was examined as a function of the cell cycle.[93,97,98] It was subsequently determined that in G0/G1 cells virtually all of the pRB exists in the underphosphorylated form. In S and G2, pRB exists largely as phosphorylated moieties.

From these data, a model was put forward to explain how pRB may suppress cell growth by acting as a cell cycle regulatory element and how T-antigen may act to overcome this activity (see Fig. 3.7). The essence of the model states that underphosphorylated pRB provides a block to exit from G1 thereby preventing cells from progressing to S-phase. This block can be removed by specific phosphorylation of pRB or complex formation of this protein with T-antigen. Once the underphosphorylated pRB is inactivated by phosphorylation or T-antigen binding, the cell can advance to the DNA replication phase of the cycle. Subsequently, time course studies capable of detecting discrete subpopulations of phosphorylated pRB suggest that pRB phosphorylation precedes entry into S-phase by several hours, and thus is not coordinated with the G1-S transition and therefore may not directly regulate it.[99] Along this same line, cell cycle specific phosphorylation of pRB has been shown to alter its affinity for the nuclear compartment of the cell,[100-103] suggesting that such tight association of the hypophoshporylated form is important for the growth regulating functions of pRB, perhaps by concentrating or sequestering factors involved in transcriptional control. Recently, a bipartite nuclear localization signal in pRB has been described, disruption of which alters its biological activity and phosphorylation in vivo.[104]

This cell cycle dependence of pRB phosphorylation raised the interesting question of the fate of T-antigen/pRB complexes during the cell cycle since T-antigen can only bind to underphosphorylated pRB. To address this, a discrete population of pRB and T-antigen was followed through the various phases of the cell cycle.[105] Cell synchronization and pulse-chase labeling studies revealed that pRB becomes phosphorylated at the beginning of S and remains so through S and G2. pRB was then completely dephosphorylated between the end of G2 and the beginning of G1. It was shown that complexes of T-antigen and underphosphorylated pRB present during G1 dissociated during S and reformed again in M or early G1 concomitant with the dephosphorylation of pRB. It has been speculated that T-antigen complexing with pRB at this point prevented an otherwise orderly progression through M, and therefore may contribute to the aneuploid condition which is commonly found in SV40-transformed cells. During the course of this investigation,[105] it was noted that newly synthesized, pulse-labeled, T-antigen could not be found complexed with underphosphorylated pRB while over time this interaction could be demonstrated. One time dependent property of T-antigen is its ability to assemble into higher order oligomers. By correlating the time

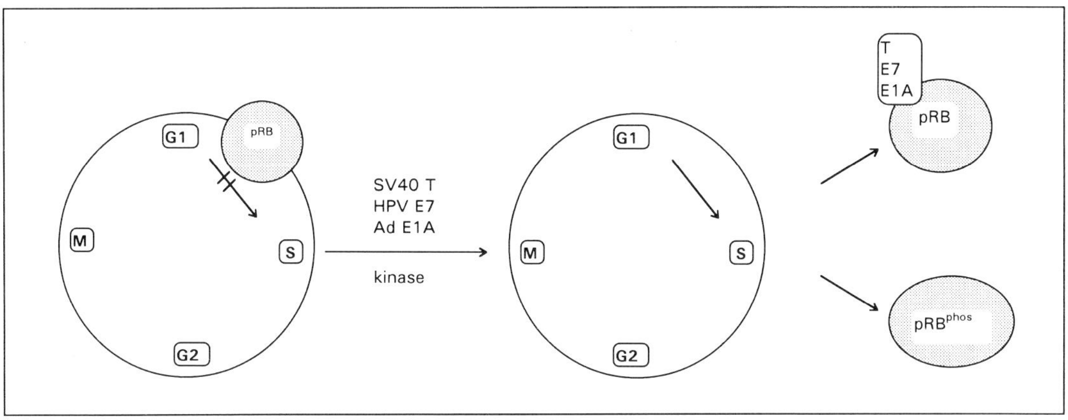

Fig. 3.7. Model for growth suppression by pRB. The underphosphorylated form of pRB appears to be active with respect to growth suppression. Underphosphorylated pRB blocks the transition from G1 into S-phase of the cell cycle. Specific phosphorylation of pRB or pRB binding to oncoproteins from DNA tumor viruses can also remove this block to cell cycle progression. Once pRB is inactivated by one of these two mechanisms, the cell can advance to the DNA replication phase of the cell cycle. Adapted from DeCaprio et al.[93]

course of T-antigen oligomerization with the appearance of pulse-labeled T-antigen complexing with underphosphorylated pRB, it appeared that in addition to an intact transforming domain sequence, T-antigen must oligomerize prior to pRB binding. Taken together with the absence of complex formation between pRB and a mutant of T-antigen which cannot oligomerize but contains an intact transforming domain, it was concluded that complex formation between T-antigen and pRB may be regulated in part by the cell cycle dependent phosphorylation and dephosphorylation of pRB and in part by the quaternary structure of T-antigen (see Fig. 3.8).

Cell cycle inhibition studies involving the growth suppressive effect of the paracrine growth factor TGF-β on T-antigen express-

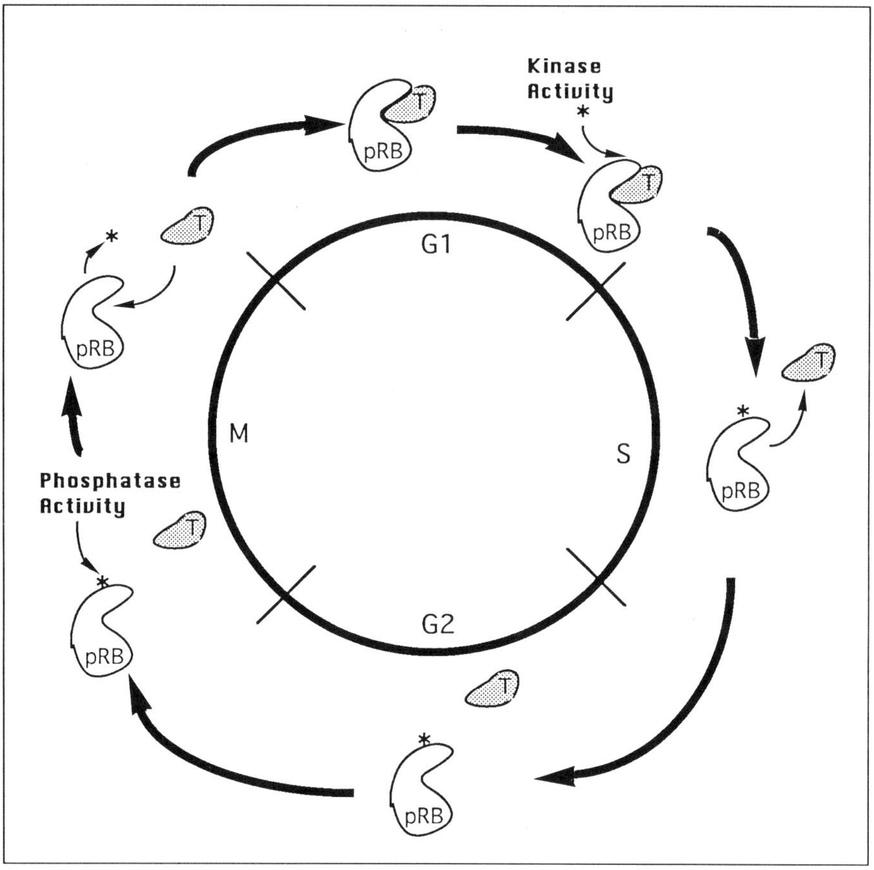

Fig. 3.8. Schematic representation of the cell cycle stage-dependent modification of pRB and association with SV40 T-antigen. Late during G1 as the cell progresses towards DNA synthesis (S-phase), pRB that is or was recently complexed with T-antigen becomes phosphorylated. Subsequently, existing pRB/T-antigen complexes are dissociated and potential complexes between these two proteins are inhibited. pRB remains phosphorylated through G2. Upon the onset of mitosis (M-phase), phosphatase activity leads to pRB dephosphorylation. Once phosphates are removed from pRB, complex formation with T-antigen can then proceed. Adapted from Ludlow.[145]

ing cells provided additional support for the hypothesis that T-antigen binding has functional consequences for pRB.[106] The addition of TGF-β to Mv1Lu mink lung epithelial fibroblasts in mid to late G1 prevents phosphorylation of pRB scheduled for this stage and arrests cells in late G1. When T-antigen was expressed in these cells, it was shown that such expression does not block the effect of TGF-β on pRB phosphorylation but greatly reduces the growth inhibitory response to TGF-β. The conclusion was that this cytokine and pRB appear to function in a common growth inhibitory pathway in which TGF-β acts to retain pRB in the underphosphorylated, growth suppressive state and that T-antigen abrogates the growth inhibitory response to TGF-β by complexing with underphosphorylated pRB thus nullifying its growth suppressive activity.

While most of the reported research on T-antigen/pRB complex formation focused on the disruption of the negative growth regulatory function of pRB, it would be surprising if T-antigen function was unaffected by binding to pRB. Indeed, it has recently been shown that pRB binding can interfere with SV40 T-antigen-dependent stimulation of DNA polymerase-α, which may be a key step for the initiation of SV40 DNA replication.[107]

6. Human papillomavirus E7 binding to pRB

E7 is a 21kDa nuclear phosphoprotein that is structurally and functionally similar to adenovirus E1A and SV40 T-antigen. In particular, the amino terminus of this protein has considerable sequence homology with the two noncontiguous transforming domain regions of E1A and the contiguous transforming domain region of T-antigen (see Fig. 3.5). The carboxy terminus contains two Cys-X-X-Cys motifs spaced 29 amino acids apart. HPV-16 E7 proteins containing mutations in this region exhibit reduced transforming, immortalizing, and transactivating efficiency, yet pRB binding is, for the most part, retained[108,109] (see Fig. 3.9). Taken together, these data suggest that these cysteine motifs are important for these functions, they do not contribute to pRB binding, nor is such binding essential for E7 to carry out immortalization.

E7 has been shown to form a complex with pRB,[110,111] preferring the underphosphorylated form of this protein[112] (see Fig. 3.10). Such binding is dependent upon the intact transforming domain region of HPV that is structurally homologous to E1A and T-antigen; mutations in this region, particularly in the amino terminal portion,[113] eliminate this binding. Mutations in this region which do eliminate pRB binding, however, do not alter the HPV-16 E7 ability to immortalize primary keratinocytes.[108] Synthetic peptides

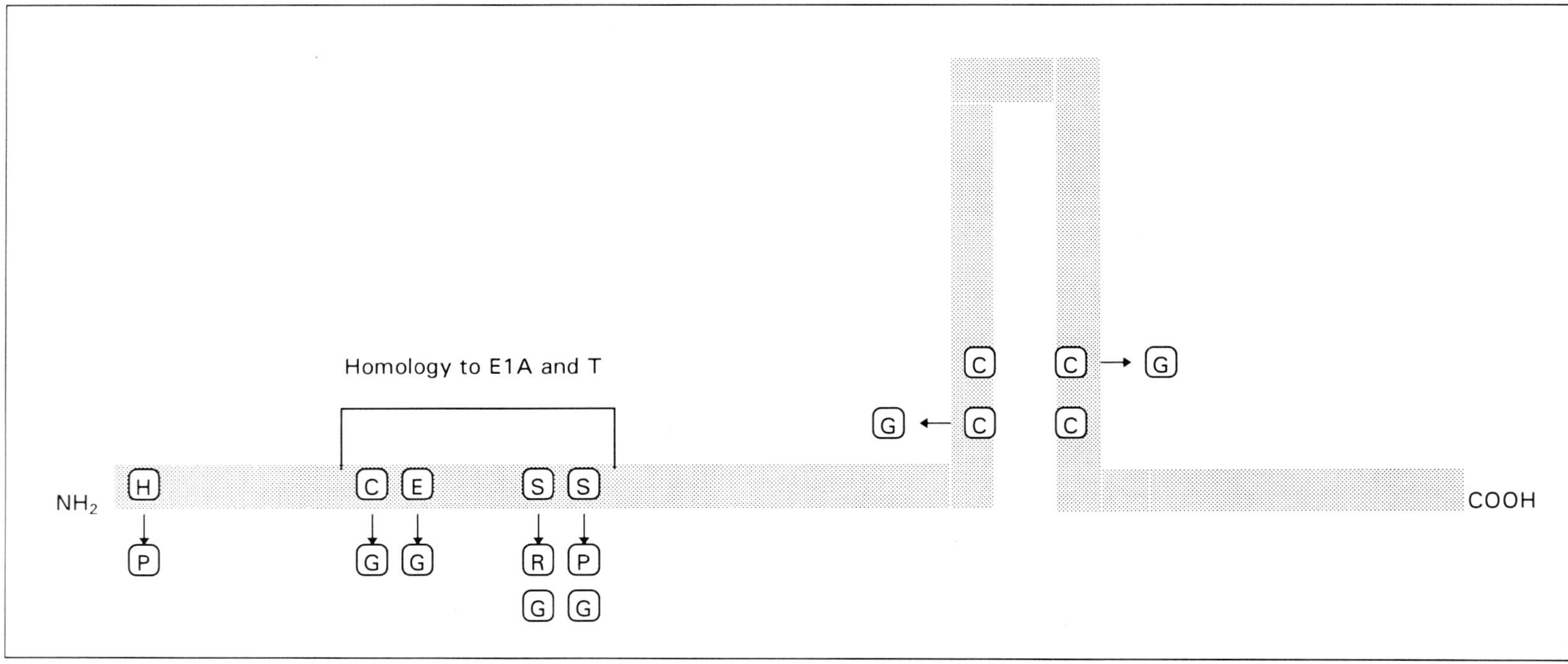

Fig. 3.9. Schematic representation of HPV E7. Depicted are the conserved cysteine motifs at the carboxy terminus and the region of similarity with Ad E1A and SV40 T-antigen. The two serine residues within this region are phosphorylated by casein kinase II. Amino acid substitutions in point mutants used to test the requirements for pRB, p107, and associated kinase activity in HPV type 16 transformation are designated by the arrows. Diagram is adapted from Davies et al.[141]

made to this region are potent antagonists to this complex formation.[114] The importance of this complex formation to the oncogenic activity of E7 is strengthened by the observation that E7 from low-risk HPV types demonstrate low affinity for pRB.[110,115] Additional structure-function analysis revealed that the transforming and transcriptional activation domains of HPV-16 E7 overlap within a region analogous to the conserved region 2 of Ad E1A, suggesting that pRB binding is necessary for both activities.[116] Deletion of sequences in conserved region 1 abrogates transformation but has minimal effect on transcriptional activation, suggesting that E7 transactivation and cellular transformation are interrelated but separable functions. A comparison of the E7 pRB binding sites revealed a consistent amino acid difference (aspartic acid/glycine) between the high-risk and low-risk viruses. Substitution mutations made at this position demonstrated that this single amino acid residue was the principle determinant responsible for the observed affinity differences between the high and low-risk HPV for pRB as well as the differences in transformation capacity. Binding of pRB does not, however, appear to be required for the immortalization property of this viral protein.[108]

It has been shown that the association of E7 with pRB inhibits the DNA binding ability of pRB.[117] Using a 60kDa version of pRB, containing the viral pocket binding domain and carboxy regions, in vitro transcription and translation, as well as expression

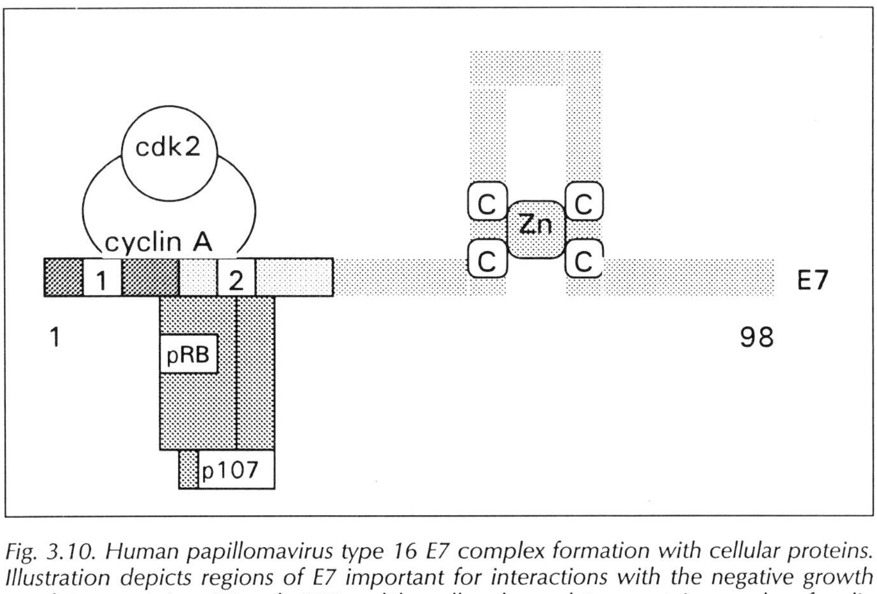

Fig. 3.10. Human papillomavirus type 16 E7 complex formation with cellular proteins. Illustration depicts regions of E7 important for interactions with the negative growth regulatory proteins pRB and p107, and the cell cycle regulatory protein complex of cyclin A and cdk2. The regions of E7 similarity with adenovirus E1A conserved region 1 and conserved region 2 (highlighted regions) necessary for pRB, p107, and cyclin A/cdk2 kinase association is also shown. Adapted from Vousden.[143]

in *E. coli*, were used to produce the protein. This protein can bind DNA cellulose quantitatively. Recombinant HPV-16 E7 protein was shown to disrupt this binding, whereas an E7 peptide that blocked binding of E2 to pRB60 restored the DNA-binding activity of this protein in the presence of E7. A point mutation in pRB60 that impairs its DNA binding ability also blocked the ability of this protein to inhibit cell growth. While it is easy to imagine how the release of cellular proteins complexed with pRB as a result of disruption by E7 could stimulate cell growth, the significance of E7 disrupting pRB DNA binding to stimulate cell growth is less clear. Perhaps the complex of proteins and DNA prevents activation of transcriptional units recognized by the complex. E7 may disrupt complexes of other proteins with pRB concurrent with pRB bound to DNA, thus activating transcription of previously suppressed genetic loci.

7. pRB Interaction With Other Large T-Antigens

The amino acid sequences of the large T-antigens of human JC virus, human BK virus, mouse polyomavirus, baboon SA12 virus, and monkey lymphotropic papovavirus are homologous to the pRB-binding sequences of SV40 T-antigen. When pRB polypeptides are synthesized in vitro and mixed with the appropriate cell lysates, stable protein complexes between pRB and the large T-antigens of all of these polyomaviruses are able to form stable complexes as detected by coimmunoprecipitation studies.[118] This suggests that these interactions have been conserved during evolution. Recently, it has been shown that polyomavirus large T-mutants which are unable to bind to pRB fail to inhibit myogenesis.[119] pRB expression appears to be necessary for myogenic activity of MyoD, since pRB-lacking cell lines cannot undergo the myogenic conversion.[120] The simplest interpretation of these data is that the physical interaction of polyoma large T-antigen with pRB is responsible for the inhibition of differentiation by removing pRB from its multiprotein complexes containing cyclin-dependent kinases, transcription factors, and other proteins with as yet unknown functions (see chapter 4). Along this same line, it has recently been demonstrated that transcription of DNA synthesis- and precursor- producing enzymes is coordinately activated in mouse cells growth arrested by serum starvation when polyoma large T-antigen is induced.[121] Mutation studies indicate that this activation requires the intact binding site for pRB and p107 within the T-antigen. The promoters of several of the genes examined carry binding sites for the E2F transcription factor, suggesting that complexes forming with this growth and cell cycle regulation factor are targets for T-antigen. Indeed, as presented in chapter 4, the currently accepted model of how viral proteins stimulate transcription is by interacting with pRB and p107, as well as additional pocket binding proteins which are capable of forming complexes with viral proteins, thus intervening and blocking the function of transcription factors such as E2F.

8. pRB and p53 Binding to EBNA-5

The herpes viruses are the largest among the group of DNA tumor viruses. Epstein-Barr virus is the most highly transforming member of this group. Encoding six nuclear proteins, three of which are required for B cell transformation, this virus is a causative agent for mononucleosis as well as fatal B cell lymphomas in immunocompromised patients, and is implicated as a causative agent in Burkitt's lymphoma and nasopharyngeal carcinoma.[122] While first discovered to colocalize with pRB in the IB4 lymphoblastoid cell line,[123] the ability of EBNA-5 to form complexes with pRB and p53 was demonstrated using bacterially produced GST-fusion proteins and EBV immortalized human lymphoblastoid cell lines.[124] Deletion mutant studies revealed a 66-amino acid peptide sufficient for this binding, while point mutations of pRB and p53 which abolish binding to other viral oncoproteins do not appear to alter binding to EBNA-5. The finding that pRB can compete for binding with p53 indicates that the interaction with pRB differs from that of the other transforming viruses. While EBNA-5 does not have an LXCXE pRB binding motif, standard for the other viral proteins that bind to pRB, it does possess a homologous region to conserved region 1 of adenovirus[125] which is necessary for pRB binding. However, while this region does not appear to be required for pRB binding, it is required for efficient B cell transformation. One possible explanation is that this region is involved in sequestering pRB at inactive sites by binding other nuclear components.[122] While the functional significance of these interactions are not known, experiments designed to test the effect of complex formation of these two tumor suppressors with EBNA-5 on the functions of pRB and p53 should address this question.

9. Adenovirus E1A Binding to Cellular Protein p300

Characterization the E1A associated proteins is likely to aid in defining the general mechanisms regulating cell growth and gene expression. One such protein is a high molecular weight protein designated p300[85] and another is the related protein termed p400.[126]

The p300 binding site is distinct from sequences involved in binding other E1A associated cellular proteins[85,127-129] (see Fig. 3.3). Characterization of this cellular protein has been greatly facilitated by the generation of specific antibodies.[130] Such biological reagents enabled researchers to determine that p300 is a ubiquitously expressed nuclear phosphoprotein that is actively phosphorylated in all phases of the cell cycle yet shows an additional phosphatase-sensitive modification specific to M-phase enriched cell populations. The polyclonal serum showed p300 structure to be sufficiently conserved in human, rat, monkey, mouse, and mink cells and to have a fairly broad tissue distribution. The conservation of this protein coupled with its presence even in resting cells suggests that it has a basal function. On the other hand, the appearance of differentially modified forms periodically throughout the cell cycle tends to support the notion that the function of p300 is modulated specifically in growing cells.

p300 binding to E1A is linked genetically to the enhancer repression function of E1A[131] and has an intrinsic DNA-binding activity demonstrating a preferential affinity for specific DNA sequences previously identified as enhancer-binding protein target sites.[132] Thus, the direct physical interaction of p300 with enhancer elements provides a biochemical basis for the genetic evidence linking the E1A-mediated enhancer repression function with the p300-binding activity of E1A. In order to address the mechanism of action, it will be important to determine whether E1A association with p300 interferes directly with p300 DNA binding. It is also possible that E1A affects different aspects of p300 function, such as potential associations between p300 and additional cellular factors. Along this line of thinking, p300 was recently shown to interact with the TATA-binding protein, possibly through intermediate protein-protein interactions.[132] p300-specific complexes and TATA-binding protein complexes include at least two additional common polypeptide species. These phosphoproteins of 64 and 59kDa are used by all three eukaryotic RNA polymerases as a component of different multisubunit complexes.[133] Taken together, these data suggest the possibility that p300, through its DNA-binding activity, modulates transcriptional activity at specific TATA box sequences and that E1A binding to p300 has some physical effect on interactions between p300, TATA-binding protein, and the TATA box.

The recent cloning of the p300 gene has greatly facilitated the structural analysis and functional characterization of its encoded protein.[134] Localized to chromosome 22q13, this protein contains three histidine and cysteine-rich regions. Using a series of templates for in vitro translation with progressive deletions, the third

Cys/His-Rich region located between amino acids 1572 and 1818 appears to be required for efficient interaction with E1A. p300 also contains a bromodomain thus implicating its ability to function as a transcriptional coactivator. Functional assays have shown that p300 can ameliorate the repressive effect of E1A on the SV40 enhancer, while molecules which cannot bind E1A can bypass E1A repression and restore activity of the enhancer, implying that p300 may function as a transcriptional adaptor protein for certain complex transcriptional regulatory elements. In this regard, it is probably essential for adenovirus to interfere with this growth suppressive function of p300 so that the host cell can enter S-phase enabling the virus to replicate its own DNA.

10. E1A BINDING TO P130

A protein having a molecular weight of 130kDa has also been found which associates with E1A[85] and may be related to p107. This interaction is dependent upon the intact nature of conserved regions 1 and 2 (see Fig. 3.3). Taking advantage of similarities between pRB and p107 to design degenerate oligonucleotides for the polymerase chain reaction to identify new gene products containing viral oncoprotein pocket-like binding structures, a new member of the retinoblastoma gene family was cloned.[135,136] Designated pRB2, this protein shows a high degree of similarity with pRB and p107, possess a pocket-like region, and binds to the transforming domain region of E1A, suggesting it too plays a role in cell growth regulation. Based on the similarity of electrophoretic mobility and molecular weight estimates of 120kDa from the cDNA sequence, pRb2 and p130 may be the same proteins. Recently, the p130 cDNA was isolated[137,138] and found to have sequence homology with the product of the retinoblastoma susceptibility gene and E1A-associated p107, but was more closely related to p107. p130 produced by in vitro translation was shown to be capable of binding to E1A while p130-specific antibody was capable of detecting this protein on immunoblots of E1A immunoprecipitates. This gene mapped to chromosome 16q12.2-13, a region that undergoes allelic loss in several types of tumors, including hepatocellular, prostate, and breast carcinomas support the idea that p130 is a tumor suppressor.

11. P107 BINDING TO VIRAL ONCOPROTEINS

The regions of E1A that are necessary for pRB association also contain the binding site for a second cellular protein, designated p107.[88] This protein can also be found complexing with SV40 and JC virus large T-antigens, and is dependent upon the intact nature of the T-antigen transforming domain,[89,139] suggesting that these associations constitute yet another component of a common mechanism for transformation shared by adenoviruses and polyomaviruses.

While the genetics of p107 interactions with the viral oncoproteins as well as other features of the protein suggests that p107 shares functional properties with pRB, they are distinct proteins; p107 is not recognized by antibodies to pRB, its peptide map is largely different from that of pRB, and this protein can be found in human retinoblastoma cell lines which lack any functional pRB DNA sequence. Upon obtaining a partial cDNA for human p107, its sequence was found to map to chromosome 20q11.2 and encode a protein of 936 amino acid residues.[128] Seven regions of p107/pRB homology were found which largely exist within the pRB pocket region that is essential for oncoprotein binding and its growth controlling function. Mapping studies using p107 deletion mutants revealed that the region of p107 between amino acids 225-816 constituted an independent T-antigen/E1A-binding domain, like the pRB pocket. It remains to be determined whether loss of p107 function is associated with the development of neoplasia, in a manner similar to pRB.

Like p53 and pRB, p107 is a likely candidate for the role of regulatory protein in T- antigen-mediated replication in vivo. Experiments supporting this used a GST-fusion protein construct, containing the 60kDa region of homology with the pRB pocket domain that binds to cellular proteins E2F, cyclins A and E, and the viral proteins T-antigen, E1A, E7, it was shown that the p107 pocket region negatively regulated T-antigen-mediated replication in vitro.[140]

Although analysis of E7 point mutants revealed a difference in the precise sequence requirement for pRB and p107 binding, the region previously identified as important for pRB binding was found to be associated with histone H1 kinase activity at the G2/M-phase of the cell cycle and binding of p107.[141] Thus, p107 binding can occur independent of pRB binding. Although pRB binding correlated with kinase activity, the restriction of this activity at the G2/M-phase implies that pRB binding may not directly mediate this activity. Since kinase defective E7 mutants are also defective in transformation, this may be an independent function of E7 which plays a role in the G2/M-phase of the cell cycle. Additionally, E7 may also associate with p107 and p130, at least as demonstrated by peptide mapping of coprecipitating cellular proteins phosphorylated by E7-associated kinase. These observations have not been confirmed by coprecipitating metabolically labeled proteins, but are supported by E7-related peptide interactions with p107 and p130.[142]

C. SUMMARY

As should be apparent from the information presented above, determining the precise role that these viral and cellular proteins play in modulating the cell cycle, and disruption of their functions during neoplasia, is a formidable task. Anyone who follows this literature knows that the biology of tumor formation is indeed complex. In studying cellular transformation by viral oncoproteins, researchers in this area have been able to put forth a working hypothesis to account for at least one underlying mechanism. Stated simply, a common feature for the transforming ability of these DNA tumor viruses involves inactivation of the growth suppressor property of one or more cellular proteins through complex formation with the viral oncoproteins. Figure 3.11 shows a summary diagram of the binding of cellular tumor suppressors to viral oncoproteins.

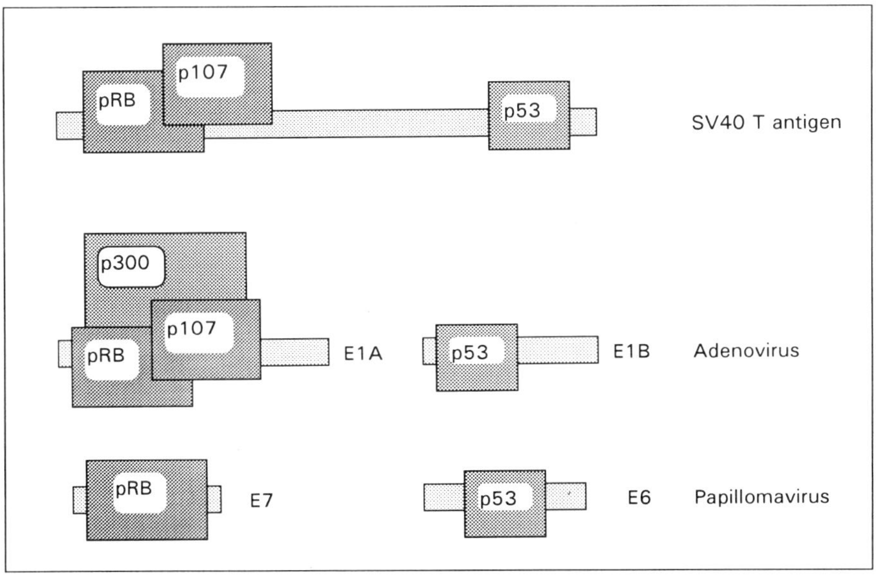

Fig. 3.11. Diagram of complex formation between cellular tumor suppressor proteins and DNA tumor virus oncoproteins. SV40 T-antigen has the ability to bind both pRB and p53 whereas two separate proteins coded for by adenovirus and papillomavirus are required to complex with these proteins. Adapted from Werness et al.[59]

References

1. Rous P. A sarcoma of the fowl transmissible by an agent separable from the tumor cells. J Exp Med 1911; 13:397-411.
2. Gross L. "Spontaneous" leukemia developing in C3H mice following inoculation, in infancy, with AK-leukemic extracts or AK-embryos. Proc Soc Exp Biol Med 1951; 76:27-32.
3. Stewart SE, Eddy BE, Gochenour AM et al. The induction of neoplasms with a substance released from mouse tumors by tissue culture. Virology 1957; 3:380-400.
4. Rigby P. The transforming genes of SV40 and polyoma. Nature 1977; 282:781-784.
5. Tooze J. Molecular virology of tumor viruses, 2nd ed, part 2, 1981. DNA tumor viruses. Cold Spring Harbor Press, New York.
6. Hassell, JA, Topp WC, Rifkin DB et al. Transformation of rat embryo fibroblasts by cloned polyoma virus DNA fragments containing only part of the early gene region. Proc Natl Acad Sci USA 1980; 77:3978-3982.
7. Mes AM, Hassell JA. Polyoma viral middle T-antigen is required for transformation. J Virol 1982; 42:621-629.
8. Triesman R, Novak V, Favoloro J et al. Transformation of host cells by an altered polyoma virus genome expressing middle T-proteins. Nature 1981; 292:595-600.
9. Rassoulzadegan M, Cowie A, Carr A et al. The roles of individual polyoma virus early proteins in oncogenic transformation. Nature 1982; 300:713-718.
10. Microbiology: Including Immunology and Molecular Genetics, 3rd edition, 1980, Harper and Row Publishers Inc., D.D. Davis, R. Dulbecco, H.N. Eisen, and H.S. Ginsberg, eds. pgs 1232-1261.
11. Tegtmeyer P. Altered patterns of protein synthesis in infection by SV40 mutants. Cold Spring Harbor Symp Quant Biol 1974; 39:9-15.
12. Prives C, Gilboa E, Revel M et al. Cell free translation of simian virus 40 early messenger RNA coding for viral T antigen. Proc Natl Acad Sci USA 1977; 74:457-461.
13. Bikel I, Montano X, Agha ME et al. SV40 small t antigen enhances the transformation activity of limiting concentrations of SV40 large T antigen. Cell 1987; 48:321-330.
14. Kalderon D, Smith AE. In vitro mutagenesis of a putative DNA binding domain of SV40 large-T. Virology 1984; 139:109-137.
15. Padgett BL, Walker DL, ZuRhein GM et al. Cultivation of papova-like virus from human brain with progressive multifocal leukoencephalopathy. Lancet 1971; 1:1257-1260.
16. Gardner SD, Field AM, Coleman DV et al. New human papovavirus (B.K.) isolated from the urine after renal transplantation. Lancet 1971; 1:1253-1257.
17. de Villiers EM. Heterogeneity of the human papillomavirus group. J Virol 1989; 63:4898-4903.
18. zur Hausen H, Schneider A. The role of papillomaviruses in anogenital cancer, p.245-263. in N.P. Salzman and P.M. Howley (ed.), The Papillomaviruses, vol. 2., 1987. The Papovaviridae. Plenum

Publishing Corp., New York.
19. Bedell MA, Jones KH, Laimins, LA. The E6-E7 region of human papillomavirus type 18 is sufficient for transformation of NIH 3T3 and rat-1 cells. J Gen Virol 1987; 61:3635-3640.
20. Pecoraro G, Morgan D, Defendi V. Differential effects of human papillomavirus type 6, 16, and 18 DNAs on immortalization and transformation of human cervical epithelial cells. Proc Natl Acad Sci USA. 1989; 86:563-567.
21. Pirisi L, Yasumoto S, Feller M et al. Transformation of human fibroblasts and keratinocytes with human papillomavirus type 16 DNA. J Virol 1987; 61:1061-1066.
22. Cullen AP, Reid R, Campion M et al. Analysis of the physical state of different human papillomavirus DNAs in intraepithelial and invasive cervical neoplasm. J Virol 1991; 65:606-612.
23. Choo K-B, Pan C-C, and Han, S-H. Integration of human papillomavirus type 16 into cellular DNA of cervical carcinoma: preferential deletion of the E2 gene and invariable retention of the long control region and the E6/E7 open reading frames. Virology 1987; 161:259-261.
24. Wilczynski SP, Pearlman L, Walker J. Identification of HPV early genes retained in cervical carcinomas. Virology 1988; 166:624-627.
25. Meyers C, Frattini MG, Hudson JB et al. Biosynthesis of human papillomavirus from a continuous cell line upon epithelial differentiation. Science 1992; 257:971-973.
26. Dollard SC, Wilson JL, Demeter LM et al. Production of human papillomavirus and modulation of the infectious program in epithelial raft cultures. Genes & Devel 1992; 6:1131-1142.
27. Houweling A, van den Elsen PJ, van der Eb AJ. Partial transformation of primary rat cells by the leftmost 4.5% fragment of adenovirus 5 DNA. Virology 1980; 105:537-550.
28. Ruley HE. Adenovirus early region 1A enables viral and cellular transforming genes to transform primary cell in culture. Nature 1983; 304:602-606.
29. van den Elsen PJ, Houweling A, van der Eb AJ. Morphological transformation of human adenovirus is determined to a large extent by gene products of region E1A. Virology 1983; 131:242-246.
30. Lane DP, Benchimol S. p53: oncogene or antioncogene ?. Genes & Devel 1991; 4:1-8.
31. Symonds H, Chen J, Van Dyke T. Complex formation between the lymphotropic papovavirus large tumor antigen and the tumor suppressor protein p53. J Virol 1991; 65:5417-5424.
32. Montenarh M. Biochemical properties of the growth suppressor/oncoprotein p53. Oncogene 1992; 7:1673-1680.
33. Meek DW, Eckhart W. Phosphorylation of p53 in normal and simian virus 40-transformed NIH 3T3 cells. Mol Cell Biol 1988; 8:461-465.
34. Tack LC, Wright JH. Altered phosphorylation of free and bound forms of monkey p53 and simian virus 40 large T antigen during lytic infection. J Virol 1992; 66:1312-1320.
35. Schmieg FI, Simmons DT. Intracellular location and kinetics of

complex formation between simian virus 40 T antigen and cellular protein p53. J Virol 1984; 52:350-355.
36. Jenkins JR, Chumakov P, Addison C et al. Two distinct regions of the murine p53 primary amino acid sequence are implicated in stable complex formation with simian virus 40 T antigen. J Virol 1988; 62:3903-3906.
37. Schmeig FI, Simmon DT. Characterization of the in vitro interaction between SV40 T antigen and p53: mapping the p53 binding site. Virology 1988; 164:132-140.
38. Tack LC, Wright JH, Deb SP et al. The p53 complex from monkey cells modulates the biochemical activities of simian virus 40 large T antigen. J Virol 1989; 63:1310-1317.
39. Tack LC, Cartwright CA, Wright JH et al. Properties of a simian virus 40 mutant T antigen substituted in the hydrophobic region: defective ATPase and oligomerization activities and altered phosphorylation accompany an inability to complex with cellular p53. J Virol 1989; 63:3362-3367.
40. Wang EH, Friedman PN, Prives C. The murine p53 protein blocks replication of SV40 DNA in vitro by inhibiting the initiation functions of SV40 large T antigen. Cell 1989; 57:379-392.
41. Deppert W, Huag M, Steinmayer T. Modulation of p53 protein expression during cellular transformation with simian virus 40. Mol Cell Biol 1987; 7:4453-4463.
42. Lin J-Y, Simmons DT. The ability of large T antigen to complex with p53 is necessary for the increased life span and partial transformation of human cells by simian virus 40. J Virol 1991; 65:6647-6653.
43. Manfredi JJ, Prives C. Binding of p53 and p105-RB is not sufficient for oncogenic transformation by a hybrid polyomavirus-simian virus 40 large T antigen. J Virol 1991; 64:5250-5259.
44. Resnick-Silverman L, Pang Z, Li G et al. Retinoblastoma protein and simian virus 40-dependent immortalization of human fibroblasts. J Virol 1991; 65:2845-2852.
45. Lin J-Y, Simmons DT. Transformation by simian virus 40 does not involve the mutational activation of p53 to an oncogenic form. Virology 1990; 176:302-305.
46. Van Roy R, Liebaut G, Mareel M et al. Partial transformation of human tumor cell lines showing defective interaction between unusual p53 gene product and SV40 large-T antigen. Oncogene 1990; 5:207-218.
47. Thompson DL, Kalderon D, Smith AE et al. Dissociation of Rb-binding and anchorage-independent growth from immortalization and tumorigenicity using SV40 mutants producing N-terminally truncated large T antigens. Virology 1990; 178:15-34.
48. Quartin RS, Cole CN, Pipas JM et al. The amino-terminal functions of the simian virus 40 large T antigen are required to overcome wild-type p53-mediated growth arrest of cells. J Virol 1994; 68:1334-1341.
49. Bargonetti J, Reynisdottir I, Friedman PN et al. Site-specific binding of wild-type p53 to cellular DNA is inhibited by SV40 T antigen and mutant p53. Genes & Devel 1992; 6:1886-1898.

50. Mietz JA, Unger T, Huibregtse JM et al. The transcriptional transactivation function of wild-type p53 is inhibited by SV40 large T-antigen and by HPV-16 E6 oncoprotein. EMBO J 1992; 11:5013-5020.
51. Michael-Michalovitz D, Yehiely F, Gottlieb E et al. Simian virus 40 can overcome the antiproliferative effect of wild-type p53 in the absence of stable large T antigen-p53 binding. J Virol 1991; 65:4160-4168.
52. Boulanger PA, Blair GE. Expression and interactions of human adenovirus oncoproteins. Biochem J 1991; 275:281-299.
53. Prives C, Manfredi JJ. The p53 tumor suppressor protein: meeting review. Genes & Devel 1993; 7:529-534.
54. Grand RJ, Lecance PS, Roberts S et al. Overexpression of p53 and c-myc in human fetal cells transformed with adenovirus early region 1. Virology 1993; 193:579-591.
55. van den Heuvel SJL, van Laar T, The I et al. Large E1B proteins of adenovirus types 5 and 12 have different effects on p53 and distinct roles in cell transformation. J Virol 1993; 67:5226-5234.
56. Yew PR, Berk AJ. Inhibition of p53 transactivation required for transformation by adenovirus early region 1B protein. Nature 1992; 357:82-85.
57. Teodoro JG, Halliday T, Whalen SG et al. Phosphorylation at the carboxy terminus of the 55-kilodalton adenovirus type 5 E1B protein regulates transforming activity. J Virol 1994; 68:776-786.
58. Debbas M, White E. Wild-type p53 mediates apoptosis by E1A, which is inhibited by E1B. Genes & Devel 1993; 7:546-554.
59. Werness BA, Levine AJ, Howley, PM Association of human papillomavirus types 16 and 18 E6 proteins with p53. Science 1990; 248:76-79.
60. Scheffner M, Werness BA, Huibregtse JM et al. The E6 oncoprotein encoded by human papillomavirus types 16 and 18 promotes the degradation of p53. Cell 1990; 63:1129-1136.
61. Huibregtse JM, Scheffner M, Howley PM. A cellular protein mediates association of p53 with the E6 oncoprotein of human papillomavirus types 16 or 18. EMBO J 1991; 13:4129-4135.
62. Huibregtse JM, Scheffner M, Howley PM. Localization of the E6-AP regions that direct human papillomavirus E6 binding, association with p53, and ubiquitination of associated proteins. Mol Cell Biol 1993; 13:4918-4927.
63. Lechner MS, Mack DH, Finicle AB et al. Human papillomavirus E6 proteins bind p53 in vivo and abrogate p53 mediated repression of transcription. EMBO J 1992; 11:3045-3052.
64. Hubbert NL, Sedman SA, Schiller JT. Human papillomavirus type 16 E6 increases the degradation rate of p53 in human keratinocytes. J Virol 1992; 66:6237-6241.
65. Blumenfeld N, Gonen H, Mayer A et al. Purification and characterization of a novel species of ubiquitin-carrier protein E2, that is involved in degradation of non-"N-end rule" protein substrates. J Biol Chem 1994; 13:9574-9581.
66. Ciechanover A, Shkedy D, Oren M et al. Degradation of the tumor

suppressor protein p53 by the ubiquitin-mediated proteolytic system requires, novel species of ubiquitin-carrier protein, E2. J Biol Chem 1994; 13:9582-9589.
67. Crook T, Tidy JA, Vousden KH. Degradation of p53 can be targeted by HPV E6 sequences distinct from those required for p53 binding and trans-activation. Cell 1991; 67:547-556.
68. Crook T, Fisher C, Masterson PJ et al. Modulation of transcriptional regulatory properties of p53 by HPV E6. Oncogene 1994; 9:1225-1230.
69. Kess TD, Slebos RJ, Nelson WG et al. Human papillomavirus 16 E6 expression disrupts the p53-mediated cellular response to DNA damage. Proc Natl Acad Sci USA 1993; 90:3988-3992.
70. Lechner MS, Mack DH, Finicle AB et al. Human papillomavirus E6 proteins bind p53 in vivo and abrogate p53-mediated repression of transcription. EMBO J 1992; 11:3045-3052.
71. Scheffner M, Munger K, Byrne JC et al. The state of the p53 and the retinoblastoma genes in human cervical carcinoma cell lines. Proc Natl Acad Sci USA 1991; 88:5523-5527.
72. Crook T, Vousden KH. Properties of p53 mutations detected in primary and secondary cervical cancers suggests mechanisms of metastasis and involvement of environmental carcinogens. EMBO J 1992; 11:3935-3940.
73. Liu F, Green MR. Promoter targeting by adenovirus E1a through interaction with different cellular DNA-binding domains. Nature 1994; 368:520-525.
74. Barbeau D, Charbonneau R, Whalen SG et al. Functional interactions within adenovirus E1A protein complexes. Oncogene 1994; 9:359-373.
75. Moran E, Mathews, MB. Multiple functional domains of the adenovirus E1A gene. Cell 1987; 48:177-178.
76. Bagchi S, Raychaudhuri P, Nevins JR. Adenovirus E1A proteins can dissociate heteromeric complexes involving E2F transcription factor: a novel mechanism of E1A *trans*-activation. Cell 1990; 62:659-669.
77. Spitkovsky D, Steiner P, Lukas J et al. Modulation of cyclin gene expression by adenovirus E1A in a cell line with E1A-dependent conditional proliferation. J Virol 1994; 68:2206-2214.
78. Cavenee WK, Dryja TP, Phillips RA et al. Expression of recessive alleles by chromosomal mechanisms in retinoblastoma. Nature 1983; 305:779-784.
79. Huang H-JS, Yee J-K, Shew J-Y et al. Suppression of the neoplastic phenotype by replacement of the RB gene in human cancer cells. Science 1988; 242:1563-1566.
80. Lee EY-HP, Chang C-Y, Hu N et al. Mice deficient for Rb are nonviable and show defects in neurogenesis and haematopoiesis. Nature 1992; 359:288-294.
81. Jacks T, Fazeli A, Schmitt EM et al. Effects of an *Rb* mutation in the mouse. Nature 1992; 359:295-300.
82. Clarke AR, Maandag ER, van Roon M et al. Requirement for a functional *Rb*-1 gene in murine development. Nature 1992; 359:328-330.
83. Ludlow JW, DeCaprio JA, Huang C-M et al. SV40 large T antigen

binds preferentially to an underphosphorylated member of the retinoblastoma susceptibility gene product family. Cell 1989; 56:57-65.
84. Yee S, Branton PE. Detection of cellular proteins associated with human adenovirus type 5 early region 1A polypeptides. Virology 1985; 147:142-153.
85. Harlow E, Whyte P, Franza R Jr et al. Association of adenovirus early region 1A proteins with cellular polypeptides. Mol Cell Biol 1986; 6:1579-1589.
86. Whyte P, Buchkovich KJ, Horowitz JM et al. Association between an oncogene and an antioncogene: the adenovirus E1A proteins bind to the retinoblastoma gene product. Nature 1988; 334:124-129.
87. Lee W-H, Lee J-Y, Shew J et al. The retinoblastoma susceptibility gene encodes a nuclear phosphoprotein associated with DNA binding activity. Nature 1987; 329:642-645.
88. Whyte P, Williamson NM, Harlow E. Cellular targets for transformation by the adenovirus E1A proteins. Cell 1989; 56:67-75.
89. Dyson N, Guida P, McCall C et al. Adenovirus E1A makes two distinct contacts with the retinoblastoma protein. J Virol 1992; 66:4606-4611.
90. Nielsch U, Fognani C, Babiss LE. Adenovirus E1A-p105(Rb) protein interactions play a direct role in the initiation but not the maintenance of the rodent cell transformed phenotype. Oncogene 1991; 6:1031-1036.
91. DeCaprio JA, Ludlow JW, Figge J et al. SV40 large tumor antigen forms a specific complex with the product of the retinoblastoma susceptibility gene. Cell 1988; 54:57-65.
92. Bartek J, Vojtesek B, Grand RJA et al. Cellular localization and T antigen binding of the retinoblastoma protein. Oncogene 1992; 7:101-108.
93. DeCaprio JA, Ludlow JW, Lynch D et al. The product of the retinoblastoma susceptibility gene has properties of a cell cycle regulatory element. Cell 1989; 58:1085-1095.
94. Ewen ME, Ludlow JW, Marsilio E et al. An amino terminal transformation governing sequence of SV40 large T antigen contributes to the binding of both $p110^{Rb}$ and a second cellular protein, p120. Cell 1989; 58:257-267.
95. Kaelin WG Jr, Ewen ME, Livingston DM. Definition of the minimal simian virus 40 large T antigen- and adenovirus E1A-binding domain in the retinoblastoma gene product. Mol Cell Biol 1990; 10:3761-3769.
96. Goodrich DW, Wang NP, Qian Y-W et al. The retinoblastoma gene product regulates progression through the G1 phase of the cell cycle. Cell 1991; 67:293-302.
97. Buchkovich K, Duffy LA, Harlow E. The retinoblastoma protein is phosphorylated during specific phases of the cell cycle. Cell 1989; 89:1097-1105.
98. Chen P-L, Scully P, Shew J-Y et al. Phosphorylation of the retinoblastoma gene product is modulated during the cell cycle and cellular differentiation. Cell 1989; 58:1193-1198.
99. Mittnacht S, Lees JA, Desai D et al. Distinct sub-populations of

the retinoblastoma protein show a distinct pattern of phosphorylation. EMBO J 1994; 13:118-127.
100. Mittnacht S, Weinberg RA. G1/S phosphorylation of the retinoblastoma protein is associated with an altered affinity for the nuclear compartment. Cell 1991; 65:381-393.
101. Templeton DJ. Nuclear binding of purified retinoblastoma gene product is determined by cell cycle-regulated phosphorylation. Mol Cell Biol 1992; 12:435-443.
102. Stokke T, Erikstein BK, Smedshammer L et al. The retinoblastoma gene product is bound in the nucleus in early G1 phase. Exp Cell Res 1993; 204:147-155.
103. Mancini MA, Shan B, Nickerson JA et al. The retinoblastoma gene product is a cell cycle-dependent, nuclear matrix-associated protein. Proc Natl Acad Sci USA 1994; 91:418-422.
104. Zacksenhaus E, Bremner R, Phillips RA et al. A bipartite nuclear localization signal in the retinoblastoma gene product and its importance for biological activity. Mol Cell Biol 1993; 13:4588-4599.
105. Ludlow JW, Shon J, Pipas JM et al. The retinoblastoma susceptibility gene product undergoes cell cycle-dependent dephosphorylation and binding to and release from SV40 large T. Cell 1990; 60:387-396.
106. Laiho M, DeCaprio JA, Ludlow JW et al. Growth inhibition by TFG-β linked to suppression of retinoblastoma protein phosphorylation. Cell 1990; 62:175-185.
107. Savoysky E, Suzuki M, Simbulan C et al. Immunopurified Rb protein inhibits SV40 T antigen-dependent stimulation of DNA polymerase α. Oncogene 1993; 8:319-325.
108. Jewers RJ, Hildebrandt P, Ludlow JW et al. Regions of human papillomavirus type 16 E7 oncoprotein required for immortalization of human keratinocytes. J Virol 1992; 66:1329-1335.
109. McIntyre M, Frattini MG, Grossman SR et al. Human papillomavirus type 18 E7 protein requires intact Cys-X-X-Cys motifs for zinc binding, dimerization, and transformation but not for Rb binding. J Virol 1993; 67:3142-3150.
110. Munger K, Werness BA, Dyson N et al. Complex formation of human papillomavirus E7 and proteins with the retinoblastoma tumor suppressor gene product. EMBO J 1989; 8:4099-4105.
111. Dyson N, Howley PM, Munger K et al. The human papilloma virus-16 E7 oncoprotein is able to bind to the retinoblastoma gene product. Science 1989; 243:934-947.
112. Imai Y, Matsushima Y, Sugimura T et al. Purification and characterization of human papillomavirus type 16 E7 protein with preferential binding capacity to the underphosphorylated form of retinoblastoma gene product. J Virol 1991; 65:4966-4972.
113. Barbosa MS, Edmonds C, Fisher C et al. The region of the HPV E7 oncoprotein homologous to adenovirus E1a and SV40 large T antigen contains separate domains for Rb binding and casein kinase II phosphorylation. EMBO J 1990; 9:153-160.
114. Jones RE, Wegrzyn RJ, Patrick DR et al. Identification of HPV-16 E7 peptides that are potent antagonists of E7 binding to the retinoblastoma suppressor protein. J Biol Chem 1990; 265:12782-12785.

115. Heck DV, Yee CL, Howley PM et al. Efficiency of binding the retinoblastoma protein correlates with the transforming capacity of the E7 oncoproteins of the human papillomaviruses. Proc Natl Acad Sci USA 1992; 89:4442-4446.
116. Phelps WC, Munger K, Yee CL et al. Structure-function analysis of the human papillomavirus type 16 E7 oncoprotein. J Virol 1992; 66:2418-2427.
117. Stirdivant SM, Hubner HE, Patric DR et al. Human papillomavirus type E7 protein inhibits DNA binding by the retinoblastoma gene product. J Virol 1992; 12:1905-1914.
118. Dyson N, Bernards R, Friend SH et al. Large T antigens of many polyomaviruses are able to form complexes with the retinoblastoma protein. J Virol 1990; 64:1353-1356.
119. Maione R, Fima GM, Holman P et al. Retinoblastoma antioncogene is involved in the inhibition of myogenesis by polyomavirus large T antigen. Cell Grow and Differ 1994; 5:231-237.
120. Gu W, Schneider JW, Condorelli G et al. Interaction of myogenic factors and the retinoblastoma protein mediates muscle cell commitment and differentiation. Cell 1993; 72:309-324.
121. Mudrak I, Ogris E, Rotheneder H et al. Coordinated *trans* activation of DNA synthesis- and precursor-producing enzymes by polyomavirus large T antigen through interaction with the retinoblastoma protein. Mol Cell Biol 1994; 14:1886-1892.
122. Wiman KG. The retinoblastoma gene: role in cell cycle control and cell differentiation. FASEB J 1993; 7:841-845.
123. Jiang W-Q, Szekely L, Wendel-Hansen V et al. Co-localization of the retinoblastoma protein in the Epstein-Barr virus-encoded nuclear antigen EBNA-5. Exp Cell Res 1991; 314-318.
124. Szekely L, Selivanova G, Magnusson KP et al. EBNA-5, an Epstein-Barr virus-encoded nuclear antigen, binds to the retinoblastoma and p53 proteins. Proc Natl Acad Sci USA 1993; 90:5455-5459.
125. Huen DS, Grand RJ, Young LS. A region of the Epstein-Barr virus nuclear antigen leader protein and adenovirus E1A are identical. Oncogene 1988; 3:729-730.
126. Egan C, Jelsma TN, Howe JA et al. Mapping of cellular protein-binding sites on the products of early-region 1A of human adenovirus type 5. Mol Cell Biol 1988; 8:3955-3959.
127. Giordano A, McCall C, Whyte P et al. Human cyclin A and the retinoblastoma protein interact with similar but distinguishable sequences in the adenovirus E1A gene product. Oncogene 1991; 6:481-486.
128. Ewen ME, Xing Y, Lawrence JB et al. Molecular cloning, chromosomal mapping, and expression of the cDNA for p107, a retinoblastoma gene product-related protein. Cell 1991; 66:1155-1164.
129. Yaciuk P, Moran E. Analysis with specific polyclonal antiserum indicates that the E1A-associated 300-kDa product is a stable nuclear phosphoprotein that undergoes cell cycle phase-specific modification. Mol Cell Biol 1991; 11:5389-5397.
130. Stein RW, Corrigan M, Yaciuk P et al. Analysis of E1A-mediated growth regulation functions: binding of the 300-kilodalton cellular

product correlates with E1A enhancer repression function and DNA synthesis-inducing activity. J Virol 1990; 64:4421-4427.
131. Rikitake Y, Moran E. DNA-binding properties of the E1A-associated 300-kilodalton protein. Mol Cell Biol 1992; 12:2826-2836.
132. Abraham SE, Lobo S, Yaciuk P et al. p300, and p300-associated proteins, are components of TATA-binding protein (TBP) complexes. Oncogene 1993; 8:1639-1647.
133. Pugh BF, Tijan RJ. Diverse transcriptional functions of the multisubunit eukaryotic TFIID complex. J Biol Chem 1992; 267:679-682.
134. Eckner R, Ewen ME, Newsome D et al. Molecular cloning and functional analysis of the adenovirus E1A-associated 300-kD protein (p300) reveals a protein with properties of a transcriptional adaptor. Genes & Devel 1994; 8:869-884.
135. Mayol X, Grana X, Baldi A et al. Cloning of a new member of the retinoblastoma gene family (pRb2) which binds to the E1A transforming domain. Oncogene 1993; 8:2561-2566.
136. Yeung RS, Bell DW, Testa JR et al. The retinoblastoma-related gene, RB2, maps to human chromosome 16q12 and rat chromosome 19. Oncogene 1993; 8:3465-3468.
137. Li Y, Graham C, Lacy S et al. The adenovirus E1A-associated 130-kD protein is encoded by a member of the retinoblastoma gene family and physically interacts with cyclins A and E. Genes & Devel 1993; 7:2366-2377.
138. Hannon GJ, Demetrick D, and Beach D. Isolation of the Rb-related p130 through its interaction with CDK2 and cyclins. Genes & Devel 1993; 7:2378-2391.
139. Dyson N, Buchkovich K, Whyte P at al. The cellular 107K protein that binds to adenovirus E1A also associates with the large T antigens of SV40 and JC virus. Cell 1989; 58:249-255.
140. Amin AA, Murakami Y, Jurwitz, J. Initiation of DNA replication by simian virus 40 T antigen is inhibited by the p107 protein. J Biol Chem 1994; 269:7735-7743.
141. Davies R, Hicks R, Crook T et al. Human papillomavirus type 16 E7 associates with a histone H1 kinase and with p107 through sequences necessary for transformation. J Virol 1993; 67:2521-2528.
142. Dyson N, Guida P, Munger K et al. Homologous sequences in adenovirus E1A and human papillomavirus E7 proteins mediate interaction with the same set of cellular proteins. J Virol 1992; 66:6893-6902.
143. Vousden K. Interactions of human papillomavirus transforming proteins with the products of tumor suppressor genes. FASEB J 1993; 7:872-879.
144. Fanning E. Simian virus 40 large T antigen: the puzzle, the pieces, and the emerging picture. J Virol 1993; 66:1289-1293.
145. Ludlow JW. Interactions between SV40 large-tumor antigen and the growth suppressor proteins pRB and p53. FASEB J 1993; 7:866-871.

CHAPTER 4

Tumor Suppressors and the Cell Cycle

A. INTRODUCTION

Yeast is the model system most often used to study the cell cycle. These unicellular fungi reproduce at a rate similar to that of bacteria and have a genome 100-fold less complex than mammalian cells. The haploid phase of their life cycle facilitates the isolation of loss-of-function mutants whose phenotype would be recessive in a diploid organism and therefore difficult to detect. As such, yeast are exceptionally useful for genetic studies of eukaryotic cell biology and have provided direct opportunities to identify, clone, and characterize the genes involved in controlling the cell cycle.

Viewed simply, eukaryotic cells reproduce themselves by duplicating their contents and then dividing in two. For most cell constituents, precise control of this duplication is not necessary; it is sufficient that the number of copies of certain molecules and organelles be approximately doubled in one cycle and that close to equal quantities of each be received by each daughter cell. The one notable exception however is the DNA, which must be duplicated exactly and divided precisely between the two daughter cells. Such precision dictates that an ordered sequence of events takes place with the understanding that progression to subsequent steps relies on the successful completion of previous ones. Over time, cells undergo both growth and division periodically. As a result, the coordination and control of these interdependent events is generally viewed as being cyclical in nature. Of paramount importance in controlling the cell cycle are two classes of proteins called cyclins and transcription factors.

B. P53 AND THE CELL CYCLE

1. P53 FUNCTIONS AS A NEGATIVE REGULATOR OF CELL GROWTH

While functional inactivation of p53 has been associated with a variety of human neoplasms, wild-type p53 has been shown to be a suppressor of neoplastic cell growth. Indeed, when a wild-type p53 gene is transfected into human colorectal carcinoma cell lines, the recipient cells formed colonies five to ten-fold less efficiently than those transfected with a mutant p53 gene.[1] Using a wild-type p53 gene under the control of a hormone-inducible promoter, induction of expression in an exponentially growing glioblastoma tumor cell line resulted in an inhibition of cell cycle progression which is reversible upon removal of the inducer or infection with SV40.[2] When growth arrested cells were stimulated to proliferate, induction of p53 expression was found to inhibit progression from G0/G1 into S-phase. While the precise mechanism of how p53 accomplishes this negative regulation of cell cycle progression is not well defined, one current hypothesis centers around the sequence-specific transactivation function of p53,[3] although results from the biological and biochemical studies of p53 mutants on the transactivating and DNA binding properties of p53 suggest that, while this property is important to its suppressor activity, this is not the only function involved in p53-mediated tumor suppression.[4]

A specific biochemical link between p53 and the cell cycle has recently been reported. Using a subtractive hybridization approach, a gene whose induction correlated with wild-type but not mutant p53 gene expression was identified in a human brain tumor cell line.[5] Termed *WAF1*, for wild-type p53-activated fragment 1, this gene was localized to chromosome 6p21.2. Introduction of cDNA encoding this gene suppressed the growth of human brain, lung, and colon tumor cells in culture. At least one p53 binding site was found to be present near *WAF1*, and this element, when placed into an artificial system with a foreign promoter, could stimulate expression of a reporter gene in the presence of wild-type p53. The authors caution, however, that whether *WAF1* is a critical target for p53 is not certain; such a binding site near *WAF1* may be coincidental and the growth inhibition mediated by WAF1 may be due to an entirely different pathway.

Following a different line of questioning, another group identified a gene they called *CiP1*, for cyclin-dependent kinases, or *cdk*, -interacting protein 1, which has a sequence identical to that of *WAF1*.[6] The product of this gene binds to cyclin complexes and inhibits the function of cyclin dependent kinases. With these data taken together, the following model for p53 function has been proposed (see Fig. 4.1). p53 expression is stimulated as a consequence of such events as DNA damage, p53 binds to *WAF1* regulatory elements and transcriptionally activates its expression. WAF1 protein will then complex with cyclin-dependent kinases thus inhibiting their activity. Preventing such phosphorylation of critical substrates will result in a blockage of cell cycle progression. Relating this to neoplastic transformation, mutant p53, which is present in the majority of these cells, cannot complex with *WAF1* thereby preventing this pathway from progressing, which will lead to unrestrained cell growth.

2. p53 Modulates Transcription of Other Cell Cycle Regulators

In view of the fact that p53 is a transcription factor functioning as a negative regulator of cell growth, the promoters of genes encoding other negative regulators may very well be targets for transcriptional regulation by p53. As a formal test of this prediction, the pRB promoter has been examined as a possible target for

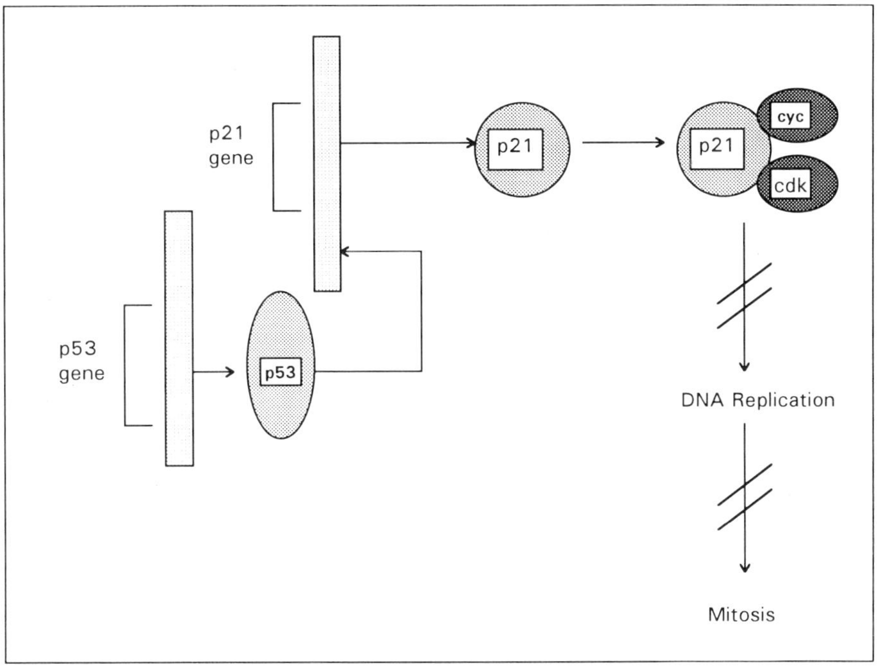

Fig. 4.1. Model of how p53 functions to suppress cell growth. Following transcription of the p53 gene, the p53 protein product acts to positively regulate transcription for p21 which, in turn, blocks the activity of cdk enzymes. This will lead to a suppression of DNA synthesis and cell division. Adapted from Marx.[142]

regulation by p53.[7] By cotransfecting a pRB promoter-luciferase reporter plasmid with varying doses of a human p53 expression plasmid into the p53(-) CCL-64 and Saos-2 cells, p53 was shown to stimulate transcription from the pRB promoter at low input doses of p53 expression plasmid, whereas transcription was slightly repressed at high input doses. The stimulatory effect was mapped between 4-92 base pairs upstream from the start of translation; repression was mapped to the basal transcriptional control region of the promoter between -207 to -185. Thus, the presence of a functional p53 binding site in the pRB promoter suggests that p53 may regulate expression of the pRB tumor suppressor gene.

Cotransfection of a rat neuroblastoma cell line with a cyclin A-luciferase expression vector and a plasmid expressing the normal or mutant p53 indicated that only the normal p53 suppresses cyclin A expression, suggesting that tumor suppressors may modulate the expression of cell cycle-regulatory genes.[8]

3. p53 Binding to the Cellular Oncoprotein MDM-2

As discussed in chapter 3, interaction of p53 with SV40 T-antigen can affect the activity of p53. In short, T-antigen binding to an evolutionarily conserved (conformational) domain of p53 inhibits its DNA binding and transcriptional activation activities. It is therefore not unreasonable to speculate that interaction of cellular proteins with this same region may also modulate the activity of p53. Using a hybrid protein composed of *Staphylococus* protein-A fused to the conformational domain of p53 (amino acids 115-295) for expression in *E. coli*, an affinity probe for p53 binding proteins was generated.[9] Probing detergent lysates of both non-small cell lung carcinoma and osteosarcoma cells, the wild-type hybrid protein specifically complexed with several major proteins having molecular weights of 45, 56, and 70kDa. This last protein is most likely a member of the heat-shock family of proteins, which have previously been reported to complex with p53, while the significance of this observation is not entirely clear.[10-12] Some additional minor species of proteins ranging in molecular weight from 30 to 90kDa were also detected. Higher levels of these binding proteins were recovered from lysates of actively growing cells compared to those from slow growing cells.

The 90kDa protein detected by this assay is most likely the product of the *mdm-2* gene, originally identified due to its amplification in a spontaneously transformed derivative of mouse BALB/c cells, which, when overexpressed, enhances the tumorigenic potential of certain cells[13] and can immortalize primary rat fibroblasts.[14] This protein has been previously shown to complex with p53 as well.[15,16]

What is the significance of MDM-2 oncoprotein binding to the tumor suppressor p53? Transactivation of the murine muscle creatine kinase enhancer by wild-type p53 in vivo has been shown to be inhibited by the MDM-2 protein, suggesting that it may interfere with the normal function of p53.[15] Molecular dissection has revealed the MDM-2 binding domain of p53 as well as the p53 binding domain on MDM-2 (see Fig. 4.2). Deletion mutants of MDM-2 incapable of complexing with p53 were also incapable of inactivating p53-mediated transactivation.[17-19] Taken together, these results underscore the importance of the p53/MDM-2 interaction in cellular growth control.

Recently, monoclonal and polyclonal antibodies directed against murine and human MDM-2 detected the 90kDa MDM-2 and at least four additional polypeptides of molecular weights 85, 76, 74, and 58-57kDa.[20] The thought is that these additional proteins arise from different spliced forms of the MDM-2 mRNA or posttranslational modifications of the protein itself. Evidence has recently been presented to support the former idea, and that splice variants differ in their ability to inhibit p53-mediated transactivation.[21] Coprecipitation studies revealed that only a subset of these proteins complex with p53, suggesting that MDM-2 may have different functions. These investigators[20] also reported on the increased MDM-2 protein level and p53/MDM-2 complex in late G1-phase, suggesting that this may be a component of the signal to enter S-phase. Recently, expression

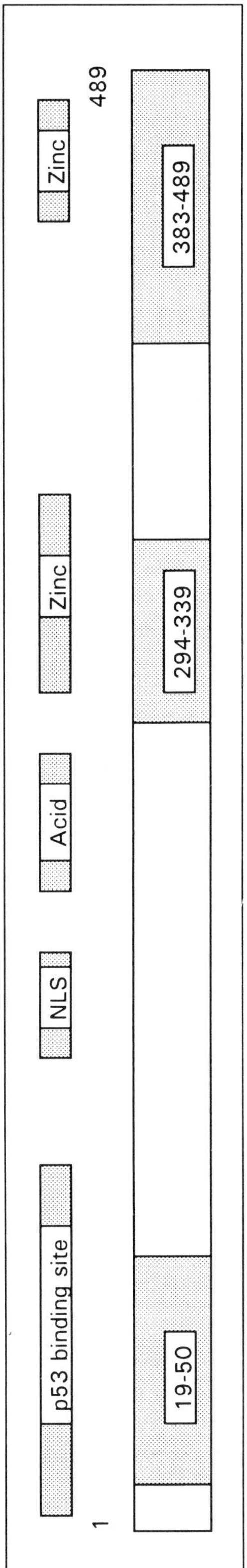

Fig. 4.2. Domain region map of mdm-2 gene product. The p53 binding region, nuclear localization signal region (NLS), acid domain (Acid), and zinc finger regions (Zinc) are indicated above the schematic with codon positions (shaded regions) for orientation. Adapted from Olson et al.[20]

of endogenous MDM-2 was enhanced by ionizing radiation at the transcriptional level in a p53-dependent manner.[22] This observation lends support to the hypothesis that MDM-2 functions in a feedback loop mechanism with p53, possibly limiting the length of p53-mediated arrest following DNA damage (see Fig. 4.3).

4. p53 Complex Formation with WT1

By studying baby rat kidney cell lines immortalized by transfection with adenovirus E1A and the Wilms' tumor gene *WT1*, coimmunoprecipitation and western blot studies revealed an association between the tumor suppressors p53 and *WT1*[23], although it has not yet been definitively proven that this interaction is a direct one. A potential functional interaction between these two tumor

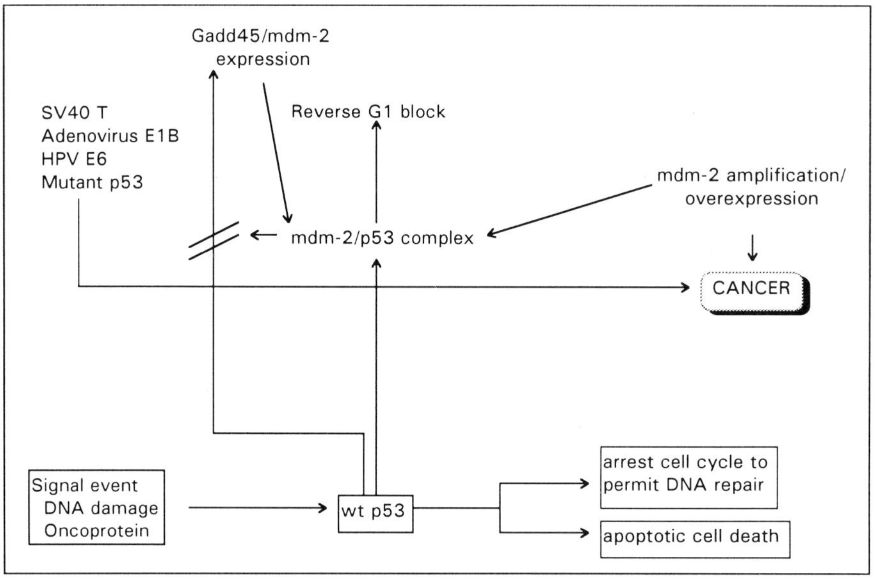

Fig. 4.3. A signal pathway model for wild-type p53 function in tumor suppression as proposed by Zambetti and Levine.[143] Signal events such as DNA damage by gamma radiation or the production of viral oncogene products stabilizes wild-type p53 protein concomitant with an increase in its steady-state level. Such an increase in wild-type p53 levels may stimulate transcription of genes involved in DNA repair and genes that negatively control cell growth. This could result in cell proliferation arrest at the G1-phase of the cell cycle. The mdm-2 *gene product, a cellular regulator which is positively responsive to elevated levels of wild-type p53, could negatively feedback on the increased levels of wild-type p53 in an autoregulatory loop pathway, which when in complex at the appropriate concentration could reverse the cell cycle block. Overexpression of the* mdm-2 *gene product appears to promote tumorigenesis. Viral oncogene products such as T-antigen, E1B, and E6 can inhibit wild-type p53 transcriptional activation and by blocking this function may contribute towards cellular transformation. Acting as a safeguard against neoplastic transformation, aberrant DNA structures or intermediates may increase wild-type p53 protein levels thus resulting in apoptosis to eliminate the cell or arrest growth to facilitate DNA repair. Loss of wild-type p53 function as a result of mutation, alteration in subcellular localization, complex formation with viral oncoproteins or negative regulators encoded by the cell may remove this safeguard against carcinogenesis.*

suppressors has been suggested by transactivation assays using their respective target sequences. It was shown that WT1 enhances the transcriptional activation of the muscle creatine kinase promoter by p53, while wild-type p53 converted WT1 from a potent transcriptional activator of the early growth response gene 1 (*EGR1*) to a transcriptional repressor. Figure 4.4 shows a functional domain map of the WT1 protein.

5. Role of p53 in Apoptosis

Apoptosis, variously referred to as programmed cell death, physiological cell death, biological cell death, or cell suicide, has been gaining widespread acceptance as we understand its importance in many areas of cell biology as part of normal cell development.[24,25] Apoptotic cells often shrink in size, loose cell-cell contact, condense their chromatin, and exhibit internucleosomal degradation of their DNA.[26] While the molecular regulation of apoptosis is far from fully understood,[27] such programmed cell death can be used as a defense against cancer[28] as well as modulating viral pathogenesis and latency.[29]

Evidence has been presented that expression of the adenovirus E1A oncoprotein can increase cellular susceptibility to apoptosis.[30,31] In addition, accumulation of p53 protein is also associated with apoptosis,[32] although not all forms of apoptosis require p53.[33-35] Induction of apoptosis by E1A is associated with deregulation of cell growth control[36] and is mediated by wild-type p53.[37] In fact, abrogation of E1A-associated apoptosis allows transformation of p53-deficient cells.[38] One other study demonstrated that ionizing radiation as well as selected chemotherapeutic agents can trigger apoptosis in E1A-expressing cells and identified the p53 molecule as required for efficient activation of the cell death program.[39] The products of the adenovirus E1B gene can inhibit p53 function either by binding directly to p53 (the 55kDa protein product) or by some indirect mechanism (the 19kDa protein product). Expression of either dominant mutant alleles of p53 or the E1B gene can block apoptosis by E1A.[37] Taken together, these results suggest that apoptosis results directly from the combined effects of E1A and p53 specifically. Alternatively, although not necessarily exclusive of this proposal, is the idea that E1A-induced apoptosis requires a block to cellular proliferation.[40] The idea here is that within an infected culture subjected to a block to proliferation, such as serum starvation or culture confluence, only some cells will experience the block immediately because some are in different growth states. E1A will induce the rest to proliferate until they

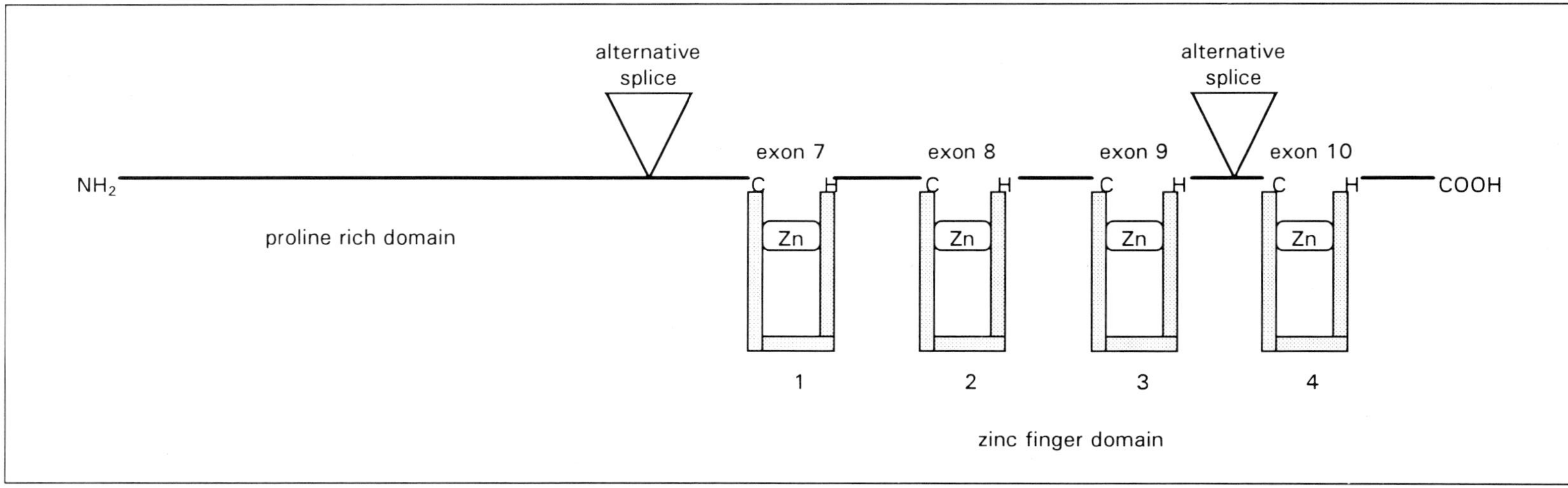

Fig. 4.4. Functional domain map of the WT1 gene product. As illustrated, the WT1 protein contains a proline-rich domain at the amino terminus and four Cys_2-His_2 zinc fingers at the carboxy terminus. Two alternative splices are evident in the gene transcript: 17 amino acids are inserted between the proline-rich domain and the first zinc finger by one alternative splice and three amino acids are inserted betwen zinc fingers 3 and 4 by the second alternative splice. Adapted from Coppes et al.[144]

also reach the block. Thus, when induction of proliferation is opposed by the proliferation block, apoptosis will occur. At the very least, these experiments argue that E1A and endogenous p53 cannot be the only factors determining apoptosis in rat cells.

A clue as to how the E1B 19kDa protein prolongs cell life may lie in its similarity to Bcl-2, a mammalian protein that blocks apoptosis.[41] In fact, the human Bcl-2 protein can suppress E1A-induced cell death just as efficiently as does the E1B 19kDa protein.[30] Recently, Bcl-2 has been shown to divert the activity of p53 from induction of apoptosis to induction of growth arrest, thus classifying Bcl-2 as a modifier of p53 function.[42] Alternatively, as suggested by Mymryk,[40] the E1B 19kDa protein may impede the passage of inhibitory growth signals to the nucleus by disrupting intermediate filaments and nuclear lamina,[43] thus avoiding the opposition between proliferation and the signals which inhibit proliferation.

C. pRB AND THE CELL CYCLE

1. pRB AS A NEGATIVE REGULATOR OF CELL GROWTH

Functional inactivation of the *pRB* gene has been observed in a variety of human tumors.[44-47] Of the naturally occurring tumors for which mutations in the *pRB* gene have been identified, at least part of the regions important for viral oncoprotein binding appear to be deficient (see Fig. 4.5). In keeping with the notion that a common feature for the transforming ability of DNA tumor viruses involves inactivation of the growth suppressor property of pRB as a result of complex formation with the viral oncoproteins (discussed in chapter 3), these data suggest that these regions of pRB are crucial for its growth suppressive activity. In contrast to these aberrant tumor cells, pRB can be found in virtually all mammalian tissues examined to date. Taken together, these observations suggest that pRB plays an important role in growth control of normal, non-neoplastic cells. Indeed, the tight regulation of phosphorylation and dephosphorylation during the cell cycle suggests that pRB itself may act as a cell cycle regulator, the inactivation of which may contribute towards unrestrained cell growth and proliferation. The ability of pRB to restrain cell cycle progression was elegantly demonstrated by microinjecting bacterially produced pRB into osteosarcoma cells lacking functional pRB.[48] It was found that microinjecting unphosphorylated pRB at least 6-10 hours before the onset of DNA replication blocked cell cycle progression at the G1-phase. Thus, a critical time window was defined in G1 during which pRB could act to impede further cycle progression.

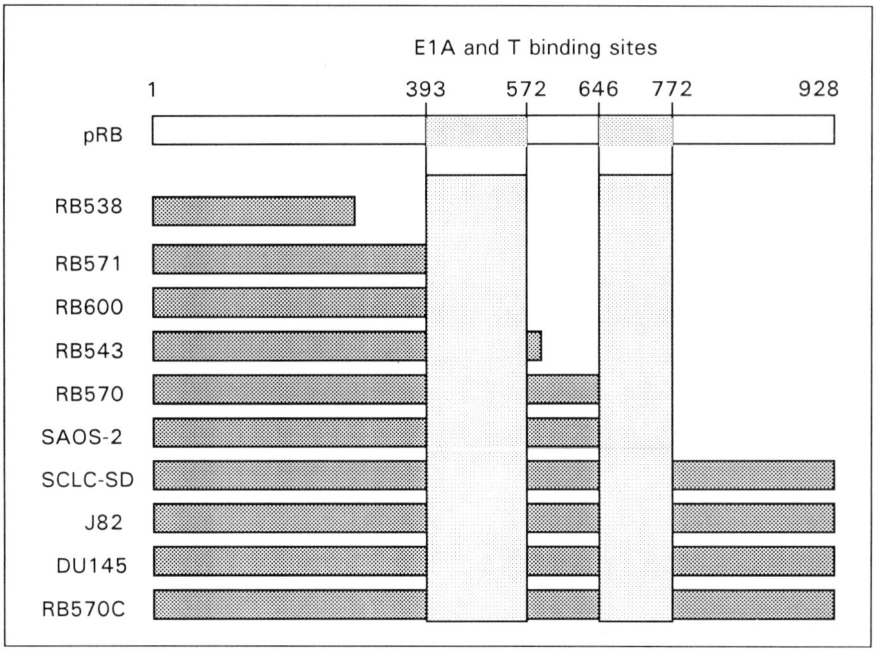

Fig. 4.5. Mutations in the pRB gene which have been identified in naturally occurring tumors. Each of these mutants is deficient in at least part of the regions of pRB deemed essential for binding to adenovirus E1A or SV40 T-antigen. Top bar represents the pRB coding region. Lightly shaded areas represent regions essential for viral oncoprotein binding. Positions of these binding regions relative to the pRB sequences present in the naturally occurring mutants (dark shaded bars) are shown. Each of these mutants contains a deletion of amino acid sequences required for E1A and T-antigen binding. Cells analyzed include those derived from retinoblastomas (RB538, RB571, RB600, RB534, RB570), an osteosarcoma (Saos-2), a small cell lung cancer (SCLC-SD), a bladder carcinoma (J82), and a prostate carcinoma (DU145). Illustration shown is an adaptation from Hu et al.[145]

To investigate the possibility that this protein may also play some role at later points during the cycle, a system of rapid inducible gene amplification to conditionally overexpress pRB during G2-phase was used.[49] Double-label fluorescence activated cell sorting (FACS) analysis showed that cells overexpressing pRB were uniformly accumulating in G2, whereas cells expressing endogenous levels were found throughout the cell cycle. These results argue in favor of the notion that pRB activity is not restricted to G1. Mechanistically, the capability of pRB to impede cell cycle progression past G2-phase may depend upon an interaction with some component(s) of the cell cycle regulatory machinery during this phase.

2. pRB Binding to the Cellular Transcription Factor E2F

Identification of the cellular targets of pRB is essential in order to define the function of this growth regulatory protein at the biochemical level. Using a series of glutathione S-transferase-pRB fusion proteins noncovalently bound to glutathione-Sepharose, at least seven cellular proteins that could bind specifically to the region of pRB which is required for interaction with the viral tumor proteins, the so-called pocket region, have been revealed.[50,51] One of these proteins is the cellular transcription factor E2F.[52-55] Originally identified as a factor involved in adenovirus E2 gene transcription, this factor is also required for efficient transcription of cellular genes involved in growth control and DNA synthesis.[56] In fact, overexpression of E2F can activate DNA synthesis in cells that would otherwise undergo growth arrest. Microinjection of E2F cDNA into quiescent cells has been shown to induce entry into S-phase.[57] Some of the genes which contain E2F-binding sites include c- and N-*myc*,[58,59] thymidine kinase,[60] DNA polymerase α,[61] dihydrofolate reductase,[62,63] *cdc2*,[64] and *RB1*.[65] For this last example, autoregulation of pRB has been found to negatively regulate expression of E2F. Thus, transcription of the pRB gene can be regulated by the transcription factor E2F, in response to the demands for pRB protein during cell cycle progression. The pRB protein can, in turn, adjust its expression level by inactivating the activity of its activator, here, E2F. Thus, in addition to phosphorylation and possible interactions with other cellular proteins, pRB regulates itself.

It is the hypophosphorylated form of pRB which has been found associated with E2F. Thus, hypophosphorylation of pRB may suppress cell growth by binding to E2F thereby effectively blocking its growth promoting activity which permits the cells to exit from G1 and enter S-phase. pRB phosphorylation would therefore release E2F from pRB and lead to transcriptional activation of growth-promoting genes (see Fig. 4.6). This does not necessarily mean, however, that binding of pRB to E2F simply inactivates E2F. Indeed, it has been shown that when pRB-E2F is bound to the E2 site, other promoter elements are inhibited thereby resulting in down-regulation of transcription.[66] These authors suggest that the E2 site alternates between a positive and a negative element as a function of the phosphorylation/dephosphorylation cycle of pRB. Thus, they suggest that pRB-E2F is an active complex that inhibits transcription and may be critical to the function of pRB; it would allow pRB to have a major effect on promoter activity by switching E2 sites from positive to negative elements. Along this same line, investigators have shown that in undifferentiated cells pRB is recruited into a complex with the viral E4-ORF6/7 product and E2F. The formation of this multiprotein complex on the E2F DNA targets correlates with the activation of

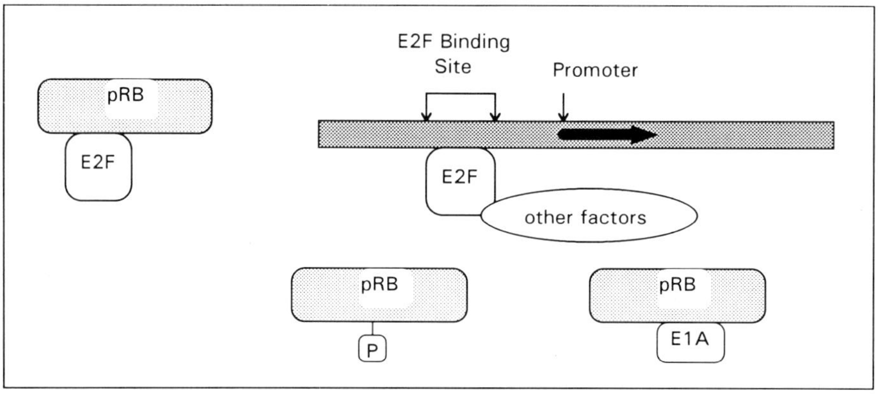

Fig. 4.6. pRB interaction with the cellular transcription factor E2F. Underphosphorylated pRB, present during G0/G1 of the cell cycle, binds to E2F thus down-regulating its transcriptional activity. E2F can be released from this complex either by phosphorylation of pRB or by viral oncoprotein binding (Ad E1A, SV40 T-antigen, HPV E7) to pRB. Upon such release, E2F can then function in concert with other factors of the transcriptional machinery as a transcriptional activator for genes containing the E2F-binding site, such as adenovirus E2, cdc2, *c-myc, DNA polymerase α, and others. Since E2F can also complex with another cellular protein known to suppress cell growth, p107, a similar model substituting p107 for pRB in complex with E2F may be proposed. Illustration shown is an adaptation from Wiman.*[146]

the viral E2a promoter. In differentiated cells, however, the activator function of pRB is transformed to a negative regulatory function.[67]

Alternatively, the DNA virus oncoproteins E1A, T-antigen, and E7 all have the ability to dissociate this complex, thus overriding the pRB-mediated transcriptional repression,[68-72] (see Fig. 4.7) which is largely dependent upon the intact nature of the conserved transformation domain sequence. Further examination of this interaction with E1A, however has revealed that independent regions of this virus protein are involved in the binding and dissociation of the E2F-protein complexes.[73,74] The data are consistent with a model whereby the initial E1A binding to the E2F-protein complex (pRB and/or p107) is facilitated by the conserved region 2 domain. Then upon normal equilibration, dissociation of pRB or p107 from E2F, the conserved region 1 domain blocks the site of interaction on pRB or p107, thus preventing the reformation of the complexes. E1A proteins lacking conserved region 1 could therefore form stable complexes with pRB-p107 and E2F, but would not release E2F. Such ability may be a common activity of these viral proteins, the purpose of which may be to stimulate quiescent cells into a proliferating state so that viral replication may proceed more effectively. In cases where lytic infection is not the end point, disruption of this complex may initiate the oncogenic process in a manner analogous to the mutation of the *RB1* gene.

Using pRB to screen expression libraries, several laboratories have been successful in isolating a cDNA clone possessing many of the properties of E2F, and designated it E2F-1.[75-77] When transfected into mammalian cells, E2F-1 was able to mediate transcription dependent upon the presence of E2F DNA binding sites. Structural analyses have shown that E2F-1 contains discrete domains

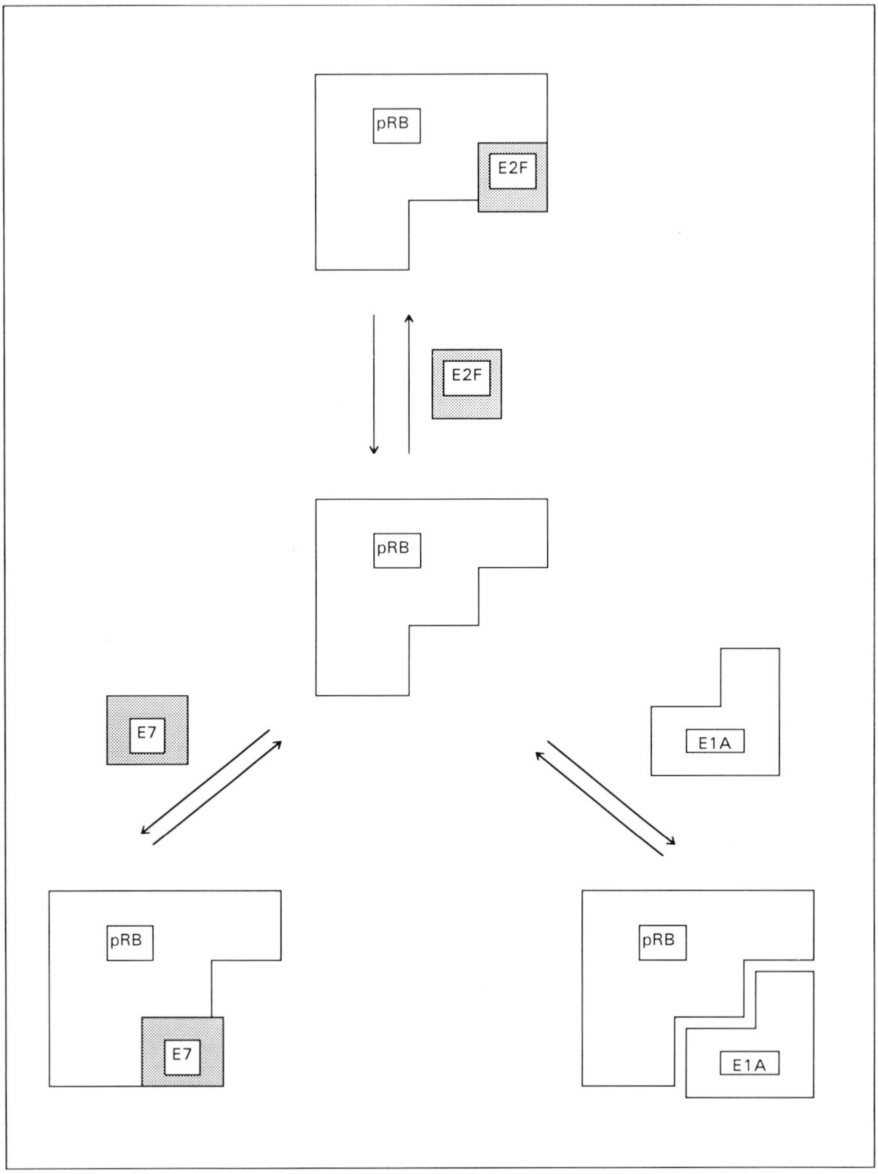

Fig. 4.7. Schematic representation of pocket-binding region of pRB interacting with human papillomavirus E7, adenovirus E1A, and E2F transcription factor. As shown, there is an overlap of the binding site on pRB for E2F and possibly E1A (through conserved region 1) but not E7. The primary interaction between E7 and possibly E1A with the pRB pocket-binding region is through the conserved region 2 binding site. As depicted, E7 inhibition of E2F binding to the pRB pocket-binding region results from steric hinderence, although other models are possible. Adapted from Huang et al.[72]

capable of mediating DNA binding, *trans*-activation, and pRB binding. In further analyzing the E2F-1 transactivation domain, pRB binding has been proposed to suppress such E2F-1 activity by virtue of direct physical association.[78] Indeed, mutations which inhibited pRB from interacting with E2F also eliminated pRB-mediated E2 promoter repression in addition to negating the growth suppressing property of pRB.[79] However, upon subsequent characterization, it was difficult to account for all of the suggested E2F activities through the function of this one protein. For example, when a detailed analysis of E2F DNA-binding activities and their cell cycle behavior in primary human T cells was performed, novel E2F complexes in addition to unbound or free E2F were resolved in gel shift experiments.[80] These results suggested that E2F is comprised of multiple, biochemically distinct DNA-binding proteins which function at different points in the cell cycle. Using low stringency hybridization, cDNA clones that encode two additional E2F-like proteins, called E2F-2 and E2F-3, were isolated.[81] Although they arise from different chromosomal locations than E2F-1, both were found to bind wild-type E2F recognition sites and pRB in vivo, and both were able to activate transcription of E2F-responsive genes in a manner dependent upon at least one functional E2F binding site.

Another protein with some of the properties of human E2F has been cloned from murine cells.[82] Designated DP-1, this protein has been shown to have some affinity for an E2F DNA-binding site, however the DNA sequence of human E2F and murine DP-1 are quite different. Recently, a mouse cDNA encoding a protein that is 86% identical to the human E2F-1 was cloned.[83] This protein can bind to and activate transcription from E2F sites. DP-1 is more abundant than E2F-1 in all stages of the cell cycle. Thus, a predominant E2F binding activity in G1-phase cells is probably due to DP-1 or, perhaps, E2F-1 interaction with constitutively expressed DP-1 via heterodimer formation. Following the cloning of the human DP-1 homolog, it was found that DP-1 and E2F-1 can associate in vitro and in vivo, and that this interaction leads to enhanced binding to E2F DNA-binding sites.[84] Such an interaction also enhances transactivation by binding of pRB,[85,86] suggesting that heterodimerization regulates E2F function thus contributing to cell cycle control. Mechanistically, DP-1 may serve as a negative regulator of a cell's exit from G1-phase until this effect is negated by phosphorylation of pRB. At this point, DP-1 may be considered a positive regulator by virtue of its ability to enhance the DNA binding activity and transactivating power of the DP-1-E2F-1 heterodimer. Taken together, these data suggest that the E2F activities described previously result from the combined action of a family of proteins.

3. Additional, and as yet, Unidentified pRB Binding Proteins

Several other proteins have also been shown to specifically complex with pRB. RBP1 and RBP2 were originally isolated and their partial cDNAs obtained based on their respective protein product's ability to associate with the pRB-pocket.[87] The full-length cDNAs encoding these proteins have recently been cloned and their respective protein products characterized.[88] Both RBP1 and RBP2 are ubiquitously expressed, although low in abundance, and rather large (approximately 200kDa). These properties contribute to the difficulty in demonstrating an association with pRB in vivo. They are both nuclear phosphoproteins possessing structural motifs suggestive of a role in transcriptional regulation. Interestingly, purified HPV E7 can dissociate this complex.

A protein having an apparent molecular weight of 45kDa has also been shown to bind to pRB, and this binding appears to be dependent upon the regions of pRB required for SV40 T-antigen binding.[51] This same group recently reported the isolation of a full length cDNA coding for p48, a protein which also forms a complex with pRB both in vitro and in vivo.[89] This protein shares sequence homology with *MS11*, a negative regulator of the Ras-cyclic AMP pathway in *S. cerevisiae* and, like *MS11*, human p48 suppresses heat-shock sensitivity in several yeast strains. While it has not been definitively proven that p48 and p46 are the same protein, they do share amino acid sequences and polyclonal antibodies raised against p48 do cross react with p46. Thus, these proteins are most likely related. Further studies are needed to ascertain the functional significance of this interaction with pRB in mammalian cells.

A 60kDa pRB binding protein, designated RBP60, has also been described.[90] It was demonstrated that additions of a glutathione-S-transferase-pRB fusion protein to purified preparations of E2F resulted in an inhibition of E2F DNA-binding. In contrast, additions to partially purified E2F preparations induced formation of a complex involving E2F and its recognition sequence. By fractionating extracts from mouse L cells, this factor has been purified and identified. Mechanistically, it has been speculated that pRB can form a binary complex with E2F which has a lower affinity for DNA compared to the ternary complex formed upon the addition of RBP60. It is likely that this protein is a D1 cyclin (see below).

4. pRB binding to Protein Phosphatase Type 1

Using a modified version of the yeast two-hybrid system to isolate human cDNAs which can encode proteins capable of binding to pRB, one clone was identified which encodes a novel type 1 serine/threonine protein phosphatase (PP1).[91] In vitro binding assays revealed that this enzyme preferentially bound to the hypophosphorylated form of pRB, and that this binding was dependent upon the sequences also required for SV40 T-antigen binding. In addition, cell synchrony experiments indicate that this association occurs from mitosis to early G1-phase, coincident with the observation that pRB exists in the hypophosphorylated state during these stages. While curious, the possibility that the enzyme binds to the phosphorylated form of pRB and catalyzes the removal of phosphates has not been excluded. Indeed, in vitro analysis has shown that PP1 is capable of dephosphorylating pRB,[92,93] and thus may be an important regulator of pRB function. This lends further support to the notion that cell cycle progression is regulated by the phosphorylation state of pRB.

5. Myc binding to pRB

pRB has also been shown to form a complex with c-myc, a cellular protein whose inappropriate expression contributes to neoplastic cell development. Using both wild-type c-*myc* and a series of mutant c-*myc* constructs to produce GST-c-myc chimeric fusion proteins, it was reported that an N-terminal domain of c-myc mediates binding of pRB.[94] HPV E7 protein could compete for this binding, suggesting that these proteins share overlapping binding sites on pRB. In addition, mutant pRB from a human tumor cell line which has a 35-amino acid deletion in its carboxy terminus did not demonstrate complex formation with c-myc. Taken together, these authors speculate that dominant and recessive oncogenes can cooperate by directly binding to each other to control progression through the cell cycle.

6. Heat-Shock Protein Interaction with pRB

pRB has also been shown to bind to the 73kDa constitutively expressed heat shock cognate protein, but not the 72kDa heat shock protein which is induced by various physiological stressors.[95] These data suggest an as yet undefined role for the cognate protein in cell growth control enacted in collaboration with pRB.

7. pRB Binding to c-abl

The nuclear c-abl tyrosine kinase, a ubiquitously expressed nonreceptor protein tyrosine kinase localized in both the cytoplasm and nucleus of resting and dividing cells, also appears to be a pRB-binding protein.[96] This interaction can be detected in several primary and established cell lines, and is mediated through a region of the carboxy terminus of pRB, outside of the pocket region important for viral oncoprotein and other cellular protein binding. Therefore, as predicted, c-abl binding to pRB is unaffected by viral oncoproteins. Such complex formation inhibits the tyrosine kinase activity of c-abl. The cell cycle dependent phosphorylation of pRB correlates with the disruption of this complex resulting in the activation of nuclear c-abl tyrosine kinase activity in S-phase cells. While the functional implications for pRB are unclear, these data suggest that pRB may supervise communication between c-abl and the cellular pocket-binding proteins, thus providing an additional level of pRB-mediated specificity for the assembly of protein complexes at specific DNA sequences. These authors propose that the protein binding domains of pRB act in concert to regulate cell growth and differentiation and have termed this the "match maker model" (see Fig. 4.8).

8. Cell Cycle Regulatory Kinase Binding to pRB

pRB has been shown to be a substrate for the cell cycle regulator cdc2 kinase.[97,98] Human cdc2 was shown to phosphorylate pRB in vitro at most of the same sites utilized in vivo. In addition, removal of cdc2 from mitotic cell extracts by immunoprecipitation results in depletion of pRB kinase activity in vitro. This finding is consistent with the observation that pRB exists in the hyperphosphorylated form at the onset of mitosis and is dephosphorylated at the completion of mitosis. A recent report suggests that, at lease in human fibroblasts, Ca^{2+}/calmodulin regulates the signaling cascade that leads to cdc2 kinase activation, pRB phosphorylation, and DNA synthesis.[99]

By employing a newly developed panel of anti-human pRB monoclonal antibodies, immunoprecipitation studies revealed that pRB can physically associate with the human $p34^{cdc2}$ or a closely related kinase.[100] In addition, preclearing cell extracts with monoclonal antibody directed against cyclin A removed virtually all of the pRB-associated kinase activity, confirming that the pRB-associated kinase contains a cyclin component, most likely cyclin A, although cyclin A does not bind directly to pRB.[101,102] Interestingly, cotransfection of human retinoblastoma cells with a $p34^{cdc2}$

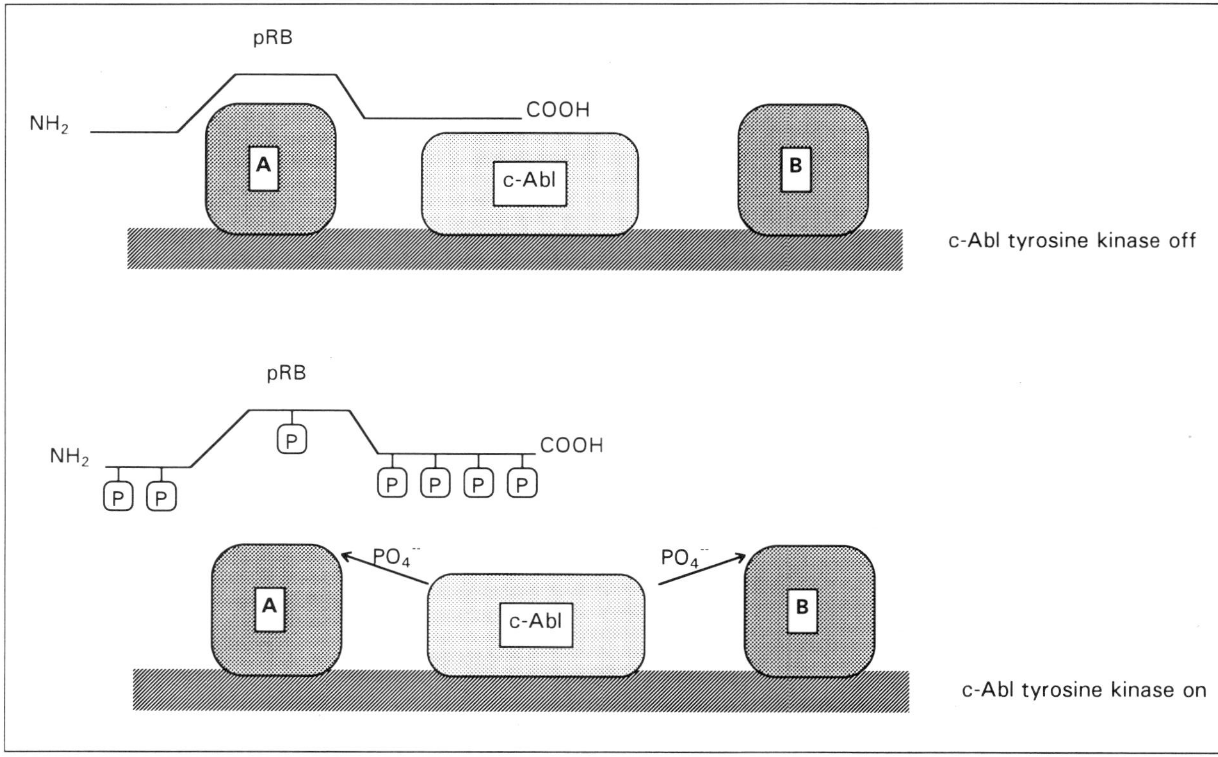

Fig. 4.8. Matchmaker model as proposed by Welch and Wang.[96] Underphosphorylated pRB forms a complex with c-Abl in G1 cells thus inhibiting the tyrosine kinase activity of c-Abl. A hypothetical complex of c-Abl, pRB, and a pRB pocket-binding protein designated A is shown. Formation of this trimolecular complex may be promoter-specific as determined by the DNA binding specificities of both c-Abl and pocket-binding protein A. Upon phosphorylation of pRB by cdk/cyclin kinases at G1/S, pRB dissociates from c-Abl and pocket binding protein A. This dissociation can lead to activation of the tyrosine kinase activity of c-Abl which in turn may phosphorylate nearby DNA binding protein B or additional DNA binding proteins thus modulating transcription.

promoter-luciferase expression vector and a plasmid expressing pRB indicated that pRB suppresses $p34^{cdc2}$ expression,[8] suggesting that this tumor suppressor modulates the expression of cell cycle regulatory genes.

A number of additional observations suggest that some additional kinase(s) may also contribute to pRB phosphorylation in vivo. The associated kinase phosphorylates most, but not all of the sites found in vivo. The most compelling evidence in support of this notion is that the initial pRB phosphorylation occurs prior to the G1/S transition, which is several hours before cyclin A is known to be expressed. Additional studies[103,104] which also employed immunoprecipitation, cell cycle arrest, phosphopeptide mapping and in vitro kinases assays further support the notion that members of the cyclin-dependent-kinase, or cdk, family, having similar substrate specificities, are involved in the phosphorylation of pRB. Of what possible biological significance could pRB binding to kinases have? First, this association may localize the kinase to pRB thereby providing specificity. Alternatively, pRB may serve to anchor the kinase so that it may be targeted to additional cellular proteins bound to pRB.

9. pRB Binding to G1 Cyclins

The cyclins consist of a family of proteins that complex with cdks which modulate crucial transitions during the cell cycle. Owing to the observation that pRB is phosphorylated during late G1 phase by a cyclin-cdk complex, the G1 cyclins, which include cyclins C, D1, D2, D3, and E, became likely candidates for a role in this interaction. It was subsequently demonstrated that both a functional and a physical interaction between pRB and the D-type cyclins exists[105,106] (see Fig. 4.9). G1 cyclins synthesized by in vitro translation were mixed with wild-type and mutant constructs of a GST-pRB fusion protein. It was determined that two of the three G1 cyclins, D2 and D3, could complex with pRB, and that this interaction was dependent upon the regions of pRB necessary for its growth suppressive activity, the same ones involved in viral oncoprotein binding. Coinfection of *Spodoptera frugiperda* (Sf9) insect cells with baculovirus vectors encoding the various D-type cyclins and cdk4 resulted in kinase activity capable of phosphorylating GST-pRB in vitro at sites identical to those phosphorylated in human T cells.[107] More recently, cdk6 has been found associated with and activated by the D-type cyclins. This kinase can phosphorylate pRB near the time when pRB phosphorylation is initiated in T cells.[108]

Interestingly cyclins D2 and D3, but not D1, functionally interact with cdk2 to phosphorylate pRB (see Fig. 4.9). Introducing

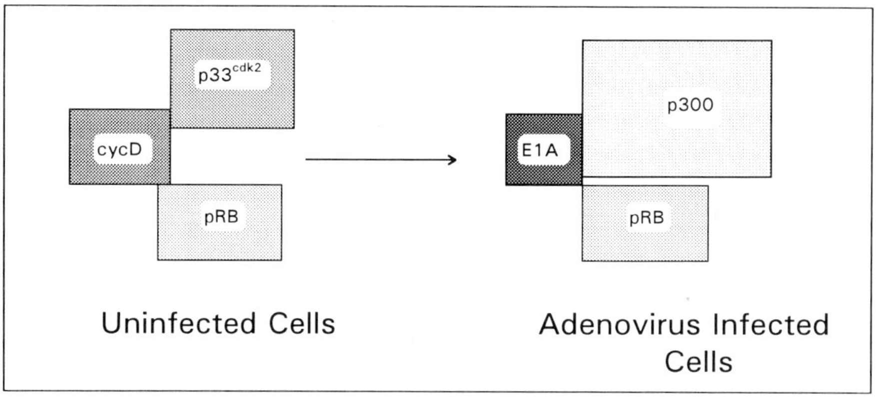

Fig. 4.9. Model of cell cycle regulatory protein complex formation in uninfected and adenovirus infected cells. In uninfected cells, cyclin D/cdk complex formation with pRB will facilitate pRB phosphorylation thus modulating its growth suppressive activity. In adenovirus infected cells, the cyclin D/cdk complex is displaced by E1A binding to pRB. Adapted from Barbeau et al.[147]

D2 and pRB into the pRB-deficient cell line Saos-2 resulted in hyperphosphorylation of pRB, inhibition of E2F binding, and reversal of the G1 block to exit the cell cycle. This was not observed upon the introduction of pRB alone or together with D1. Thus, the D-type cyclins display different functions and may also play a role as both regulatory subunits of cdk4 and as adaptor proteins that physically target active enzyme complexes to particular substrates.

These cyclins all contain an LXCXE sequence motif, which most likely plays a role in their binding to pRB. Dowdy and coworkers[106] report that D1 and D3, not D2 as reported by Ewen,[105] complex with pRB. Both groups report that cyclin D1 promoted pRB phosphorylation inefficiently and weakly overrode an pRB-induced G1 block in Saos-2 cells. Dowdy[106] however found that a D1 mutant defective in the LXCXE-binding motif, and therefore unable to bind pRB was more effective than the wild-type protein in overriding this G1 block. Ewen,[105] on the other hand, found an analogous mutation in D2 that had the opposite effect; the G1 block was not overridden. Thus Dowdy[106] suggests that pRB has a negative effect on the growth promoting function of D1; the observed weakness of D1 in overcoming a G1 block and lack of pRB phosphorylation support this. Ewen,[105] however, may argue that the opposite biological effects occur upon D2 and pRB presentation to Saos-2 cells. In summary, although D1 can enhance pRB phosphorylation in vitro,[107] there is conflicting evidence concerning the actual mechanism in vivo.[105,106] However, it has been reported that cotransfection of D1 with pRB can partially override the cell growth inhibition of pRB alone in an pRB-deficient cell line.[109]

In a study of human esophageal carcinoma cell lines and primary tumors, it was found that those in which D1 was overexpressed had normal levels of pRB, while those that had no pRB had low levels of D1 cyclin expression.[110] These results suggest that the suppressive effect of pRB on cell cycle progression may be alleviated during tumor development either by the loss of pRB expression or by increased cyclin D1 expression.

D. pRB AND TRANSCRIPTIONAL REGULATION

A 30 base pair element within the *c-fos* promoter has been identified as the target of transcriptional regulation by pRB.[111] Termed the RCE (retinoblastoma control element), three nuclear proteins called RCPs (retinoblastoma control proteins) have been shown to complex with this promoter element in vitro.[112] RB1 does not appear to encode the RCPs. A single RCP binding site for the c-*fos* RCE has been identified and similar sequences are found in the promoters for c-*myc* and *TGF-β1*, both of which are regulated by pRB. RCPs were also shown to interact with these sequences. One of the nuclear proteins that binds to these RCEs in vitro is the transcription factor Sp-1. Sp-1 also stimulates RCE-dependent transcription in vivo.[113] Thus, pRB may regulate transcription in part by virtue of its ability to functionally interact with Sp-1.

pRB can also regulate expression of other TGF-β isoforms. Using chimeric *TGF-β1* promoter-chloramphenicol acetyltransferase gene constructs, pRB induced TGF-β1 expression in mink lung and human adenocarcinoma cells and can also repress its activity in two different mouse cell lines.[114] These results suggest a cell type dependence for human pRB to regulate TGF-β1 expression either negatively or positively. pRB has also been shown to activate transcription of the human *TGF-β2* gene. The promoter element responsible for this pRB-mediated transcriptional regulation is a binding site for ATF-2, a member of a transcription factor family.[115] Thus, pRB may also constrain cellular proliferation by activating expression of a growth inhibitory factor, TGF-β2.

The *neu* oncogene encodes a protein that is believed to function as a growth factor receptor. Amplification and overexpression of this gene (also known as *erb*B2 or *HER2*) has been shown to be involved in human breast and lung cancers.[116] Inactivation of pRB has also been implicated as a contributing factor in these neoplasms. Studies aimed at ascertaining whether pRB can antagonize neu-induced transformation revealed that pRB can repress *neu* oncogene transcription by targeting a novel enhancer in the *neu* promoter.[117] These results have identified neu as the first growth factor receptor that is a target for pRB, implicating pRB as regulator of the expression of genes of different functions. It will be interesting to see whether pRB can function as a tumor suppressor in breast and lung cancers by repressing *neu* expression.

Using several breast carcinoma cell lines, evidence has recently been presented that IFN-γ induction of several class II major histocompatibility genes is pRB dependent.[118] Although the data do not permit one to unequivocally state that pRB is functioning as a transactivator of these genes, the data strongly suggest that pRB represents the first transactivator identified that is required for the coordinate induction of the class II genes by IFN-γ.

E. p107 AND THE CELL CYCLE

The biological and biochemical function of human p107 has been greatly facilitated by the isolation of a full length cDNA clone.[119] Following transfection into a variety of human tumor cell lines, a colony forming assay, similar to that which has been used for pRB and p53, revealed that overexpression of p107 can inhibit proliferation. However this effect is cell type dependent. Cells that were sensitive to p107 overexpression were blocked during G1-phase of the cycle, a condition which could be overcome by E1A expression. It is interesting to note that E1A rescue of the G1 block in p107-transfected cells was more efficient than in pRB-transfected cells. In addition, rescue of p107-mediated growth arrest was examined by coexpressing several cyclin proteins which have been implicated in modulating the function of both pRB and p107. Differences were observed here also. It was found that cell cycle arrest by p107 could not be efficiently rescued by coexpression of either cyclin A or E. Previous experiments involving the ectopic expression of cyclin A or E detected the hyperphosphorylation of pRB, which presumably inactivates its function, thus rescuing these transfected cells. These results suggest that p107 has a similar, but not identical, role in growth suppression as pRB.

F. p107 BINDING TO E2F

E2F has also been shown to independently interact with the pRB-related protein p107.[120-122] E2F/p107 complexes can be distinguished by virtue of whether they are associated with the cell cycle kinase cyclin A/cdk2 or cyclin E/cdk2.[120-124] Following the development of a panel of monoclonal antibodies, coimmunoprecipitation studies revealed that p107 can complex with additional polypeptides having many of the properties of E2F, yet they were distinct from the E2F found associated with pRB.[125] The genetic analysis of a series of mutant E2F-1 cDNAs revealed that the sequences required for interactions with pRB and p107 do indeed overlap, each requiring the intact carboxy-terminal acidic domain of E2F-1, thus providing a structural basis for the observed mutually exclusive interactions.[126] Taken together, these data strengthen the argument that, rather than being redundant or equivalent regulators, pRB and p107 provide different aspects of E2F regulation. Such p107-containing complexes are cell cycle-stage regulated, with their appearance correlating with the formation of the active kinase complexes.[122,123,127] While the role of these complexes is not as well defined as those involving pRB, p107 can inhibit E2F-mediated transactivation and it has been demonstrated that the p107-E2F complexes are targets for the oncoproteins of several DNA tumor viruses.[119,128,129] These data suggest that p107 and cell cycle-dependent kinases act together to modulate E2F activation thereby influencing transcription of E2F-responsive genes at precise stages of the cell cycle.

G. p107 INTERACTION WITH MYC

By screening a cDNA expression library with a Myc fusion protein, it has recently been shown that p107 can interact with the amino-terminal transactivating domain of Myc.[130] Subsequently p107 was shown to interact in vivo and suppress the transactivating function of Myc. Interestingly, mutant forms of Myc containing alterations in the transactivating domain were found to be resistant to p107-mediated suppression. While the exact mechanism for this p107-mediated suppression of Myc transactivation remains to be determined, these data suggest that disruption of a possible regulatory interaction between Myc and p107 may be important in both cell cycle control as well as tumorigenesis.

H. p300 AND THE CELL CYCLE

In view of their results implicating p300 as a transcriptional adaptor protein, Eckner[131] has proposed that one function of p300 in cell cycle inhibition is to activate, in concert with cell-type specific transcription factors, enhancers of genes required for arrest in G0/G1 as well as terminal differentiation (see chapter 3).

I. CYCLIN A – TARGETS FOR TUMOR SUPPRESSORS AND VIRAL ONCOPROTEINS

Cyclin A can bind to the p107 pocket, but not the pRB pocket.[101,102] This binding appears to be dependent upon the unique spacer region in p107 separating the two related collinear subsegments (A and B) common to both pRB and p107 (see Fig. 4.10). The existence of a complex suggests a direct link between cell cycle control and p107 function.

Cyclin A also complexes with DRTF1 which facilitates the efficient assembly of pRB into the complex.[69] This suggests a role for cyclin A in regulating transcription and defines a transcription factor through which molecules which regulate the cell cycle in a negative fashion, such as pRB, and in a positive fashion, such as cyclin A, interact.

Using rat lung epithelial cells immortalized by SV40 T-antigen, recent experiments have revealed that this viral oncoprotein can increase p34[cdc2] expression and cyclin A, while pRB binding is not essential for this effect.[132] Low levels of T-antigen which cannot support cell proliferation are able to maintain p34[cdc2] and cyclin A at nearly maximal levels. Such an elevation appears to be a direct consequence of T-antigen expression rather than an outcome of unrelated events that occurred during immortalization of the parental primary cells. Similar findings were previously reported for adenovirus E1A, which demonstrated that

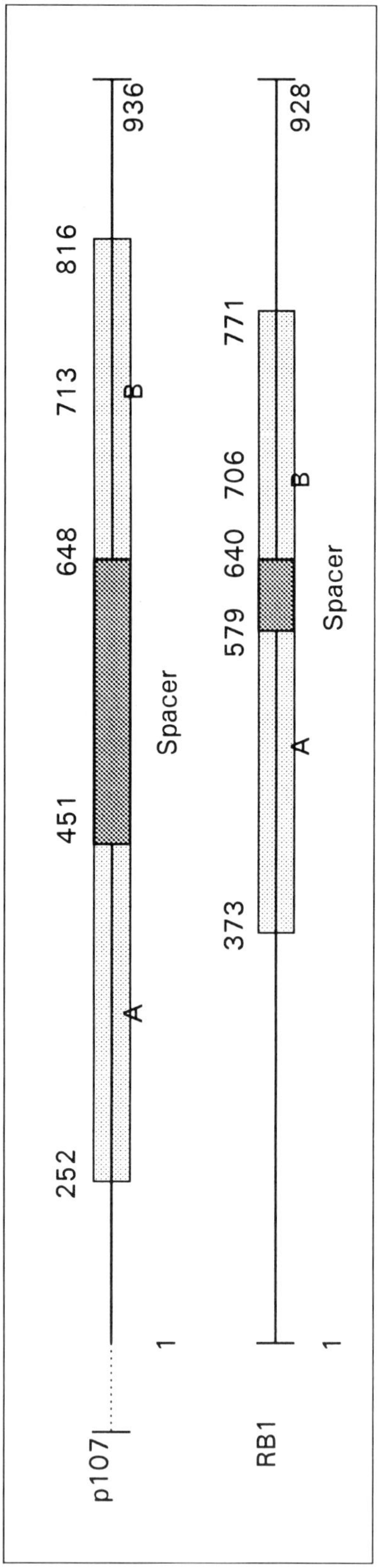

Fig. 4.10. Illustration depicting the major elements of similarity between p107 and pRB, which lie within subsegments of their respective pocket-binding regions, indicated by the thick line. Subsegments are designated A and B, with the amino acid residues indicated numerically. The spacer region between the two A and B subsegments, in addition to being of different length, also contain different amino acid sequences. Adapted from Ewen at al.[101]

induction of pRB phosphorylation was correlated with E1A-mediated induction of $p34^{cdc2}$ expression and kinase activity.[133] This observation is consistent with the possibility that $p34^{cdc2}$ is the pRB kinase.[133]

Immunoprecipitation studies have shown that T-antigen can associate with cyclin A, $p33^{cdk2}$, and histone H1 kinase activity.[134] The association of cyclin A appears to be independent of p53 association, as determined by experiments in which cell extracts were depleted of p53. Since depletion of pRB and p107 was not accomplished, it is possible that what is detected is the quaternary complexes containing cyclin A, T-antigen, cdk, and either pRB or p107. While the association between T-antigen, cyclin A, and $p33^{cdk2}$ is stable, only a minor fraction of either participant is involved in these complexes, suggesting that the interaction is indirect and requires a third or fourth molecular participant. Both pRB and p107 can bind cyclin A and $p33^{cdk2}$ independently of viral oncoproteins[100,102] (see Fig. 4.11). Likewise, similar arguments have previously been put forth as a possible explanation for cyclin A binding to E1A[135,136] and, most recently, cyclin E-$p33^{cdk2}$ binding to E1A.[127]

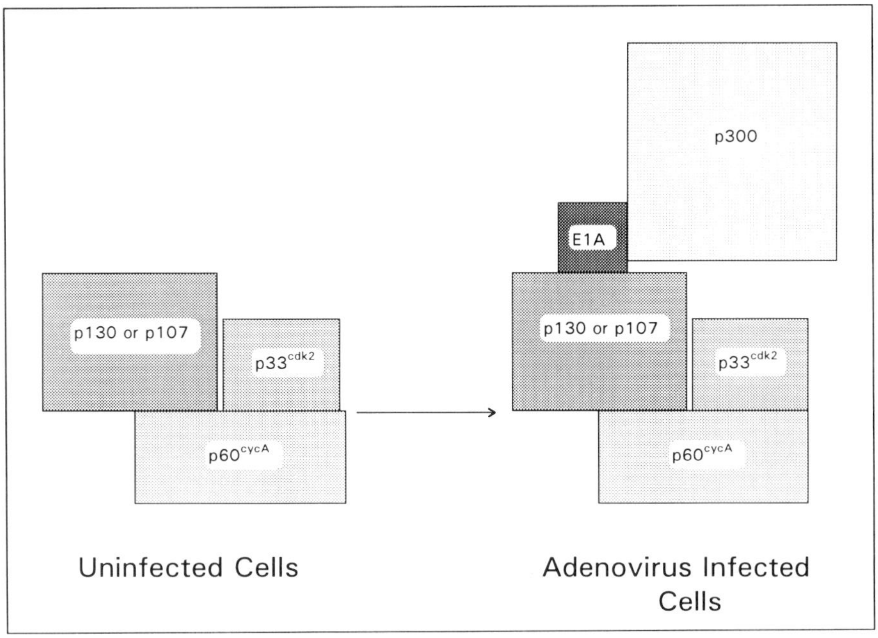

Fig. 4.11. Model of cell cycle regulatory protein complex formation in uninfected and adenovirus infected cells. In uninfected cells, cyclin A/cdk complex will bind to p107 and p130 separately, thus facilitating the phosphorylation of these proteins. Such modification modulates the functions of these proteins. In adenovirus infected cells, E1A joins this complex through its interaction with p107 or p130, and brings to this complex the p300 protein. In this scenario, the role of E1A in regulating the cell cycle may be to control the biological activity of p107, p130, and p300 by enhancing their phosphorylation by cyclin A/cdk. Adapted from Barbeau et al.[147]

The process of defining how tumor suppressors and viral oncoproteins modulate the cell cycle may also be assisted by our understanding of the specificities and activities of the enzymes involved in modifying their respective protein structure. T-antigen, p53, and pRB are all posttranslationally modified by phosphorylation, a process which affects their functions. The appearance and disappearance of phosphate groups on these proteins appears to be related to specific phases of the cell cycle, suggesting that the timing of this modification is critical to cell proliferation. While much work has been done in characterizing the kinases responsible for adding phosphates to these proteins, investigations into the specific phosphatases that remove these groups, while existent[93,137-139] are comparatively pauce. Phosphatase activity on T-antigen and pRB may be an especially fruitful area for investigation since dephosphorylation of these proteins has been shown to influence the ability to replicate SV40 DNA in vitro and block cells from exiting G1. By understanding the specificities and activities of the enzymes involved in dephosphorylating these proteins, a wealth of information regarding the cellular mechanisms involved in normal cell growth control, and the aberration of these mechanisms during neoplastic transformation, may be gained.

J. APC TUMOR SUPPRESSOR ASSOCIATION WITH CATENINS

Mutations in the *APC* gene have been implicated in the progression of colorectal tumorigenesis. In attempts to understand the function of this protein in this regard, a specific antibody to APC was generated and coimmunoprecipitation studies were carried out to determine whether APC can associate with other cellular proteins. Those studies revealed that APC can form a complex with the E-cadherin-associated proteins α- and β-catenin.[140,141] Cadherins are cell surface molecules that are important for morphogenesis. Binding of catenins is essential for cadherin function perhaps by anchoring them to the cytoskeleton. These data suggest an important link between tumor initiation and cell adhesion. Perhaps APC modulates the interaction between cadherins and catenins affecting pathways through which intercellular interactions control cell growth and differentiation.

K. SUMMARY

As should be apparent, the so-called tumor suppressor proteins function not only to slow or halt the growth of cancerous cells, but they also function to negatively regulate the growth of non-cancerous or normal cells. That regulation of cell cycle progression appears to be at the level of transcription whereby the suppressors act directly as transcription factors or indirectly by complexing with other transcription factors thereby modulating their activity. Modification of the tumor suppressor proteins by phosphorylation, as a consequence of cell division cycle-dependent kinases, appears to play a pivotal role in directing their ability to function in this capacity. Thus, by virtue of the observations that the tumor suppressors demonstrate cell cycle dependent changes in activity and/or in the extent of their modification, it may be suggested that the negative growth regulatory proteins be included in the class of cellular proteins known as cyclins.

References

1. Baker SJ, Markowitz S, Fearon ER et al. Suppression of human colorectal carcinoma cell growth by wild-type p53. Science 1990; 249:912-915.
2. Mercer WE, Shields MT, Amin M et al. Negative growth regulation in a glioblastoma tumor cell line that conditionally expresses human wild-type p53. Proc Natl Acad Sci USA 1990; 87:6160-6170.
3. Vogelstein B, Kinzler KW. p53 function and dysfunction. Cell 1992; 70:523-526.
4. Unger T, Mietz JA, Scheffner M et al. Functional domains of wild-type and mutant p53 proteins involved in transcriptional regulation, transdominant inhibition, and transformation suppression. Mol Cell Biol 1993; 13:5186-5194.
5. El-Deiry WS, Tokino T, Velculescu VE et al. *WAF1*, a potent mediator of p53 tumor suppression. Cell 1993; 75:817-825.
6. Harper JW, Adami GR, Wei N et al. The p21 cdk-interacting protein Cip1 is a potent inhibitor of G1-cyclin-dependent kinases. Cell 1993; 75:805-816.
7. Osifchin NE, Jiang D, Fujita-Ohtani N et al. Identification of a p53 binding site in the human retinoblastoma susceptibility gene promoter. J Biol Chem 1994; 269:6383-6389.
8. Yamamoto M, Yoshida M, Ono K et al. Effect of tumor suppressors on cell cycle-regulatory genes: RB suppresses $p34^{cdc2}$ expression and normal p53 suppresses cyclin A expression. Exp Cell Res 1994; 210:94-101.
9. Maxwell SA, Roth JA. Binding of cellular proteins to a conformational domain of tumor suppressor protein p53. Oncogene 1993; 8:3421-3426.
10. Hinds, PW, Finlay CA, Frey AB et al. Immunological evidence for the association of p53 with a heat shock protein, hsc70, in p53-plus-*ras*-transformed cell lines. Mol Cell Biol 1987; 7:2863-2869.
11. Finlay CA, Hinds PW, Tan T-H et al. Activating mutations for transformation by p53 produce a gene product that forms an hsc70-p53 complex with an altered half-life. Mol Cell Biol 1988; 8:531-539.

12. Sturzbecher H-W, Addison C, Jenkins JR. Characterization of mutant p53-hsp72/73 protein-protein complexes by transient expression in monkey Cos cells. Mol Cell Biol 1988; 8:3740-3747.
13. Fakharzadeh S, Trusko RS, George DL. Tumorigenic potential associated with enhanced expression of a gene that is amplified in a mouse tumor cell line. EMBO J 1991; 10:1565-1569.
14. Finlay CA. The *mdm-2* oncogene can overcome wild-type p53 suppression of transformed cell growth. Mol Cell Biol 1993; 13:301-306.
15. Momand J, Zambetti GP, Olsen DC et al. The *mdm-2* oncogene product forms a complex with the p53 protein and inhibits p53-mediated transactivation. Cell 1992; 69:1237-1245.
16. Oliner JD, Kinzler KW, Meltzer PS et al. Amplification of a gene encoding a p53-associated protein in human sarcomas. Nature 1992; 358:80-83.
17. Brown DR, Deb S, Munoz RM et al. The tumor suppressor p53 and the oncoprotein simian virus 40 T antigen bind to overlapping domains on the MDM2 protein. Mol Cell Biol 1993; 13:6849-6857.
18. Chen J, Marechal V, Levine AJ. Mapping of the p53 and mdm-2 interaction domains. Mol Cell Biol 13:4107-4114.
19. Oliner JD, Pietenpol JA, Thiagalingam S et al. Oncoprotein MDM2 conceals the activation domain of tumor suppressor p53. Nature 1993; 362:857-860.
20. Olson DC, Marechal V, Momand J et al. Identification and characterization of multiple mdm-2 proteins and mdm-2-p53 protein complexes. Oncogene 1993; 8:2352-2360.
21. Haines DS, Landers JE, Engle LJ et al. Physical and functional interaction between wild-type p53 and mdm2 proteins. Mol Cell Biol 1994; 14:1171-1178.
22. Chen C-Y, Oliner JD, Zhan Q et al. Interactions between p53 and MDM2 in a mammalian cell cycle checkpoint pathway. Proc Natl Acad Sci USA 1994; 91:2684-2688.
23. Maheswaran S, Park S, Bernard A et al. Physical and functional interaction between WT1 and p53 proteins. Proc Natl Acad Sci USA 1993; 90:5100-5104.
24. Raff MC. Social controls on cell survival and cell death. Nature 1992; 356:398-400.
25. Taylor R. Cell death makes for lively research. J NIH Res 1993; 5:59-62.
26. Wyllie AH. Glucocorticoid-induced thymocyte apoptosis is associated with endogenous endonuclease activation. Nature 1980; 284:555-556.
27. Williams GT, Smith CA. Molecular regulation of apoptosis: genetic controls on cell death. Cell 1993; 74:777-779.
28. Williams G. Programmed cell death: apoptosis and oncogenesis. Cell 1991; 65:1097-1098.
29. White E and Gooding LR. (1994) Regulation of apoptosis by human adenoviruses, p111-141. in L.D. Tomei and F.O. Cope (ed.), Apoptosis: the molecular basis for cell death II. Cold Spring Harbor

Laboratory Press, Cold Spring Harbor, N.Y.
30. Rao L, Debbas M, Sabbatini P et al. The adenovirus E1A proteins induce apoptosis which is inhibited by the E1B 19K and Bcl-2 proteins. Proc Natl Acad Sci USA 1992; 89:7742-7746.
31. White E, Sabbatini P, Debbas M et al. The 19-kilodalton adenovirus E1B transforming protein inhibits programmed cell death and prevents cytolysis by tumor necrosis factor α. Mol Cell biol 1992; 12:2570- 2580.
32. Lowe S, Ruley HE. Stabilization of the p53 tumor suppressor is induced by adenovirus-5 E1A and accompanies apoptosis. Genes & Devel 1993; 7:535-545.
33. Lowe SW, Schmitt EM, Smith SW et al. p53 is required for radiation-induced apoptosis in mouse thymocytes. Nature 1993; 362:847-849.
34. Clarke AR, Purdie CA, Harrison DJ et al. Thymocyte apoptosis induced by p53-dependent and independent pathways. Nature 1993; 362:849-852.
35. Berges RR, Furuya Y, Remington L et al. Cell proliferation, DNA repair, and p53 function are not required for programmed death of prostatic glandular cells induced by androgen ablation. Proc Natl Acad Sci USA 1993; 90:8910-8914.
36. White E, Cipriani P, Sabbatini P et al. Adenovirus E1B 19-kilodalton protein overcomes the cytotoxicity of E1A proteins. J Virol 1991; 65:2968-2978.
37. Debbas M, White E. Wild-type p53 mediates apoptosis by E1A which is inhibited by E1B. Genes & Devel 1993; 7:546-554.
38. Lowe SW, Jacks T, Housman DE et al. Abrogation of oncogene-associated apoptosis allows transformation of p53-deficient cells. Proc Natl Acad Sci USA 1994; 91:2026-2030.
39. Lowe SW, Ruley HE, Jacks T et al. p53-dependent apoptosis modulates the cytoxicity of anticancer agents. Cell 1993; 74:957-967.
40. Mymryk JS, Shire K, Bayley ST. Induction of apoptosis by adenovirus type 5 E1A in rat cells requires a proliferation block. Oncogene 1994; 9:1187-1193.
41. Hockenbery D, Nunez G, Milliman C et al. Bcl-2 is an inner mitochondrial membrane protein that blocks programmed cell death. Nature 1990; 348:334-336.
42. Chiou S-K, Rao L, White E. Bcl-2 blocks p53-dependent apoptosis. Mol Cell Biol 1994; 14:2556-2563.
43. White E, Cipriani R. Role of adenovirus E1B proteins in transformation: altered organization of intermediate filaments in transformed cells that express the 19-kilodalton protein. Mol Cell Biol 1990; 10:120-130.
44. Lee EY-HP, To H, Shew J-Y et al. Inactivation of the retinoblastoma susceptibility gene in human breast cancers. Science 1988; 241:218-221.
45. Horowitz JM, Park S-H, Bogenmann E et al. Frequent inactivation of the retinoblastoma anti-oncogene is restricted to a subset of human tumor cells. Proc Natl Acad Sci USA 1990; 87:2775-2779.
46. Yokota J, Akiyama T, Fung Y-K et al. Altered expression of the

retinoblastoma (RB) gene in small-cell carcinoma of the lung. Oncogene 1988; 3:471-475.
47. Venter DJT, Bevan KL, Ludwig RL et al. Retinoblastoma gene deletions in human glioblastomas. Oncogene 1991; 6:445-448.
48. Goodrich DW, Wang NP, Quian, Y-W et al. The retinoblastoma gene product regulates progression through the G1 phase of the cell cycle. Cell 1991; 67:293-302.
49. Karantza V, Maroo A, Fay D et al. Overproduction of Rb protein after the G1/S boundary causes G2 arrest. Mol Cell Biol 1993; 13:6640-6652.
50. Kaelin WG, Pallas DC, DeCaprio JA et al. Identification of cellular proteins that can interact specifically with the T/E1A-binding region of the retinoblastoma gene product. Cell 1991; 64:521-532.
51. Huang S, Lee W-H, Lee EY-HP. A cellular protein that competes with SV40 T antigen for binding the retinoblastoma gene product. Nature 1991; 350:160-162.
52. Bagchi S, Weinmann R, Raychaudhuri P. The retinoblastoma protein copurifies with E2F-1, an E1A-regulated inhibitor of the transcription factor E2F. Cell 1991; 65:1063-1072.
53. Bandera LR, La Thangue NB. Adenovirus E1A prevents the retinoblastoma gene product from complexing with a cellular transcription factor. Nature 1991; 351:494-497.
54. Chellappan SP, Hiebert S, Mudryi M et al. The E2F transcription factor is a cellular target for the RB protein. Cell 1991; 65:1053-1061.
55. Chittenden T, Livingston DM, Kaelin WG. The T/E1A-binding domain of the retinoblastoma gene product can interact selectively with a sequence-specific DNA-binding protein. Cell 1991; 65:1073-1082.
56. Nevins J. E2F: a link between the Rb tumor suppressor protein and viral oncoproteins. Science 1992; 258:424-429.
57. Johnson DG, Schwarz JK, Cress WG et al. Expression of transcription factor E2F1 induces quiescent cells to enter S phase. Nature 1993; 365:349-352.
58. Hiebert SW, Blake M, Azizkhan J et al. Role of E2F transcription factor in E1A-mediated *trans*-activation of cellular genes. J Virol 1991; 65:3547-3552.
59. Heibert SW, Lipp M, Nevins JR. E1A-dependent trans-activation of the human MYC promoter is mediated by the E2F factor. Proc Natl Acad Sci USA 1989; 86:3594-3598.
60. Kim YK, Lee AS. Identification of a 70-base-pair cell cycle regulatory unit within the promoter of the human thymidine kinase gene and its interaction with cellular factors. Mol Cell Biol 1991; 11:2296-2302.
61. Pearson BE, Nasheuer HP, Wang TS. Human DNA polymerase alpha gene: sequences controlling expression in cycling and serum-stimulated cells. Mol Cell Biol 1991; 11:2081-2095.
62. Slansky J, Li Y, Kaelin WG et al. A protein synthesis- dependent increase in E2F1 mRNA correlates with growth regulation of the dihydrofolate reductase promoter. Mol Cell Biol 1993; 13:1610-1618.
63. Blake MC, Azizkhan JC. Transcription factor E2F is required for

efficient expression of the hamster dihydrofolate reductase gene in vitro and in vivo. Mol Cell Biol 1989; 9:4994-5002.
64. Dalton S. Cell cycle regulation of the human cdc2 gene. EMBO J 1992; 11:1797-1804.
65. Shan B, Chang C-Y, Jones D et al. The transcription factor E2F-1 mediates the autoregulation of RB gene expression. Mol Cell Biol 1994; 14:299-309.
66. Weintraub SJ, Prater CH, Dean DC. Retinoblastoma protein switches the E2F site from positive to negative element. Nature 1992; 358:259-262.
67. Bocco JL, Reimund B, Chatton B et al. Rb may act as a transcriptional co-activator in undifferentiated F9 cells. Oncogene 1993; 8:2977-2986.
68. Chellappan S, Kraus VB, Kroger B et al. Adenovirus E1A, simian virus 40 tumor antigen, and human papillomavirus E7 protein share the capacity to disrupt the interaction between transcription factor E2F and the retinoblastoma gene product. Proc Natl Acad Sci USA 1992; 89:4549-4553.
69. Bandara LR, Adamczewski JP, Hunt T et al. Cyclin A and the retinoblastoma gene product complex with a common transcription factor. Nature 1991; 352:249-251.
70. Zamanian M, La Thangue NB. Adenovirus E1a prevents the retinoblastoma gene product from repressing the activity of a cellular transcription factor. EMBO J 1992; 11:2603-2610.
71. Pagano M, Durst M, Joswig S et al. Binding of the human E2F transcription factor to the retinoblastoma protein but not to cyclin A is abolished in HPV-16-immortalized cells. Oncogene 1992; 7:1681-1686.
72. Huang PS, Patrick DR, Edwards G et al. Protein domains governing interactions between E2F, the retinoblastoma gene product, and human papillomavirus type 16 E7 protein. Mol Cell Biol 1993; 13:953-960.
73. Ikeda M-A, Nevins JR. Identification of distinct roles for separate E1A domains in disruption of E2F complexes. Mol Cell Biol 1993; 13:7029-7035.
74. Fattaey AR, Harlow E, Helin K. Independent regions of adenovirus E1A are required for binding to and dissociation of E2F-protein complexes. Mol Cell Biol 1993; 13:7267-7277.
75. Helin K, Wu CL, Lees JA et al. A cDNA encoding a pRB-binding protein with properties of the transcription factor E2F. Cell 1992; 70:337-350.
76. Kaelin WG, Krek W, Sellers WR et al. Expression cloning of a cDNA encoding a retinoblastoma-binding protein with E2F-like properties. Cell 1992; 70:351-364.
77. Shan B, Zhu X, Chen P-L et al. Molecular cloning of cellular genes encoding retinoblastoma-associated proteins: identification of a gene with properties of the transcription factor E2F. Mol Cell Biol 1992; 12:5620-5631.
78. Flemington EK, Speck SH, Kaelin WG Jr. E2F-1-mediated transactivation is inhibited by complex formation with the

retinoblastoma susceptibility gene product. Proc Natl Acad Sci USA 1993; 90:6914-6918.

79. Hiebert SW. Regions of the retinoblastoma gene product required for its interaction with the E2F transcription factor are necessary for E2 promoter repression and pRb-mediated growth suppression. Mol Cell Biol 1993; 13:3384-3391.
80. Chittenden T, Livingston DM, DeCaprio, JA. Cell cycle analysis of E2F in primary human T cells reveals novel E2F complexes and biochemically distinct forms of free E2F. Mol Cell Biol 1993; 13:3975-3983.
81. Lees JA, Saito M, Vidal M et al. The retinoblastoma protein binds to a family of E2F transcription factors. Mol Cell Biol 1993; 13:7813-7825.
82. Girling R, Partridge JF, Bandera LR et al. A new component of the transcription factor DRTF1/E2F. Nature 1993; 362:83-87.
83. Li Y, Slansky JE, Myers DJ et al. Cloning, chromosomal location, and characterization of mouse *E2F1*. Mol Cell Biol 1994; 14:1861-1869.
84. Helin K, Wu C-L, Fattaey AR et al. Heterodimerization of the transcription factors E2F-1 and DP-1 leads to cooperative transactivation. Genes & Devel 1993; 7:1850-1861.
85. Helin K, Harlow E, Fattaey AR. Inhibition of E2F-1 transactivation by direct binding of the retinoblastoma protein. Mol Cell Biol 1993; 13:6501-6508.
86. Krek W, Livingston DM, Shirodkar S. Binding to DNA and the retinoblastoma gene product promoted by complex formation of different E2F family members. Science 1992; 262:1557-1560.
87. Defeo-Jones D, Huang PS, Jones RE et al. Cloning of cDNA for cellular proteins that bind to the retinoblastoma gene product. Nature 1991; 352:251-254.
88. Fattaey AR, Helin K, Dembski MS et al. Characterization of the retinoblastoma binding proteins RBP1 and RBP2. Oncogene 1993; 8:3149-3156.
89. Qian Y-W, Wang Y-CJ, Hollingsworth RE et al. A retinoblastoma-binding protein related to a negative regulator of Ras in yeast. Nature 1993; 364:648-652.
90. Ray SK, Arroyo M, Bagchi S et al. Identification of a 60-kilodalton RB-binding protein, RBP60, that allows the RB-E2F complex to bind DNA. Mol Cell Biol 1992; 12:4327-4333.
91. Durfee T, Becherer K, Chen PL et al. The retinoblastoma protein associates with the protein phosphatase type 1 catalytic subunit. Genes & Devel 1993; 7:555-569.
92. Alberts AS, Thornburn AM, Shenolikar S et al. Regulation of cell cycle progression and the nuclear affinity of the retinoblastoma protein by protein phosphatases. Proc Natl Acad Sci USA 1993; 90:388-392.
93. Ludlow JW, Glendening CL, Livingston DM et al. Specific enzymatic dephosphorylation of the retinoblastoma protein. Mol Cell Biol 1993; 13:367-372.
94. Rustgi AK, Dyson N, Bernards R. Amino-terminal domains of *c-myc*

and *N-myc* proteins mediate binding to the retinoblastoma gene product. Nature 1991; 352:541-544.
95. Nihei T, Takahashi S, Sagae S et al. Protein interaction of retinoblastoma gene product pRb110 with M_r 73,000 heat shock cognate protein. Cancer Res 1993; 53:1702-1705.
96. Welch PJ, Wang JYJ. A C-terminal protein-binding domain in the retinoblastoma protein regulates nuclear c-Abl tyrosine kinase in the cell cycle. Cell 1993; 75:779-790.
97. Lin BT-Y, Gruenwald S, Morla AO et al. Retinoblastoma cancer suppressor gene product is a substrate of the cell cycle regulator cdc2 kinase. EMBO J 1991; 10:857-864.
98. Lees JA, Buchkovich KJ, Marshak DR et al. The retinoblastoma protein is phosphorylated on multiple sites by human cdc2. EMBO J 1991; 10:4279-4290.
99. Takuwa N, Zhou W, Kumada M et al. Ca^{2+}-dependent stimulation of retinoblastoma gene product phosphorylation and p34cdc2 kinase activation in serum-stimulated human fibroblasts. J Biol Chem 1994; 268:138-145.
100. Hu Q, Lees JA, Buchkovich KJ et al. The retinoblastoma protein physically associates with the human cdc2 kinase. Mol Cell Biol 1992; 12:971-980.
101. Ewen ME, Faha B, Harlow E et al. Interaction of p107 with cyclin A independent of complex formation with viral oncoproteins. Science 1992; 255:85-87.
102. Faha B, Ewen ME, Tsai L-H et al. Interaction between human cyclin A and adenovirus E1A-associated p107 protein. Science 1992; 255:87-90.
103. Kitagawa M, Saitoh S, Ogino H et al. cdc2-like kinase is associated with the retinoblastoma protein. Oncogene 1992; 7:1067-1074.
104. Akiyama T, Ohuchi T, Sumida S et al. Phosphorylation of the retinoblastoma protein by cdk2. Proc Natl Acad Sci USA 1992; 89:7900-7904.
105. Ewen ME, Sluss HK, Sherr CJ et al. Functional interactions of the retinoblastoma protein with mammalian D-type cyclins. Cell 1993; 73:487-497.
106. Dowdy SF, Hinds PW, Louie K et al. Physical interaction of the retinoblastoma protein with human D cyclins. Cell 1993; 73:499-511.
107. Kato J-Y, Matsushime H, Hiebert SW et al. Direct binding of cyclin D to the retinoblastoma gene product (pRb) and pRb phosphorylation by the cyclin D-dependent kinase CDK4. Genes & Devel 1993; 7:331-342.
108. Meyerson M, Harlow E. Identification of G1 kinase activity for cdk6, a novel cyclin D partner. Mol Cell Biol 1994; 14:2077-2086.
109. Hinds PW, Mittnacht S, Dulic V et al. Regulation of the retinoblastoma protein functions by ectopic expression of human cyclins. Cell 1992; 70:993-1006.
110. Jiang W, Zhang Y-J, Kahn SM et al. Altered expression of cyclin D1 and retinoblastoma genes in human esophageal cancer. Proc Natl Acad Sci USA 1993; 90:9026-9030.

111. Robbins PD, Horowitz JM, Mulligan RC. Negative regulation of human *c-fos* expression by the retinoblastoma gene product. Nature 1990; 346:668-671.
112. Udvadia AJ, Rogers KT, Horowitz JM. A common set of nuclear factors bind to promoter elements regulated by the retinoblastoma protein. Cell Gro Diff 1992; 3:597-608.
113. Udvada AJ, Rogers KT, Higgins PDR et al. Sp-1 binds promoter elements regulated by the RB protein and Sp-1-mediated transcription is stimulated by RB coexpression. Proc Natl Acad Sci USA 1993; 90:3265-3269.
114. Kim S-J, Lee H-D, Robbins PD et al. Regulation of transforming growth factor β1 gene expression by the product of the retinoblastoma-susceptibility gene. Proc Natl Acad Sci USA 1991; 88:3052-3056.
115. Kim S-J, Wagner S, Liu F et al. Retinoblastoma gene product activates expression of the human *TGF-β2* gene through transcription factor ATF-2. Nature 1993; 358:331-334.
116. Dougall WC, Qian X, Greene MI. Interaction of the Neu/p185 and EGF receptor tyrosine kinases: Implications for cellular transformation and tumor therapy. J Cell Biochem 1993; 53:61-73.
117. Yu D, Matin A, Hung M-C. The retinoblastoma gene product suppresses *neu* oncogene-induced transformation via transcriptional repression of *neu*. J Biol Chem 1992; 267:10203-10206.
118. Lu Y, Ussery GD, Muncaster M et al. Evidence for retinoblastoma protein (RB) dependent and independent IFN-γ responses: RB coordinately rescues IFN-γ induction of MHC class II gene transcription in noninducible breast carcinoma cells. Oncogene 1994; 9:1015-1019.
119. Zhu L, van den Heuvel S, Helin K et al. Inhibition of cell proliferation by p107, a relative of the retinoblastoma protein. Genes & Devel 1993; 7:1111-1125.
120. Cao L, Faha B, Dembski M et al. Independent binding of the retinoblastoma protein and p107 to the transcription factor E2F. Nature 1992; 355:176-179.
121. Devoto SH, Mudryi M, Pines J et al. A cyclin A-specific protein kinase complex possess sequence-specific DNA binding activity: p33cdk2 is a component of the E2F-cyclin A complex. Cell 1992; 68:167-176.
122. Shirodkar S, Ewen M, DeCaprio JA et al. The transcription factor E2F interacts with the retinoblastoma product and a p107-cyclin A complex in a cell cycle-regulated manner. Cell 1992; 68:157-166.
123. Lees E, Faha B, Dulic V et al. Cyclin E/cdk2 and cyclin A/cdk2 kinases associate with p107 and E2F in a temporally distinct manner. Genes & Devel 1992; 6:1874-1885.
124. Mudryj M, Devoto SH, Hiebert SW. Cell cycle regulation of E2F transcription factor involves an interaction with cyclin A. Cell 1991; 65:1243-1253.
125. Dyson N, Dembski M, Fattaey A et al. Analysis of p107-associated proteins: p107 associates with a form of E2F that differs from pRB-associated E2F-1. J Virol 1993; 67:7641-7647.
126. Cress WG, Johnson DG, Nevins JR. A genetic analysis of the E2F1

gene distinguishes regulation by RB, p107, and adenovirus E4. Mol Cell Biol 1993; 13:6314-6325.

127. Faha B, Harlow E, Lees E. The adenovirus E1A-associated kinase consists of cyclin E-p33^{cdk2} and cyclin A-p33^{cdk2}. J Virol 1993; 67:2456-2465.
128. Schwarz JK, Devoto SH, Smith EJ et al. Interactions of the p107 and Rb proteins with E2F during the cell proliferation response. EMBO J 12:1013-1020.
129. Zamanian M, La Thangue NB. Transcriptional repression by the Rb-related protein p107. Mol Biol Cell 1993; 4:389-396.
130. Gu W, Bhatia K, Magrath IT et al. Binding and suppression of the Myc transcriptional activation domain by p107. Science 1994; 264:251-254.
131. Eckner R, Ewen ME, Newsome D et al. Molecular cloning and functional analysis of the adenovirus E1A-associated 300-kD protein (p300) reveals a protein with properties of a transcriptional adaptor. Genes & Devel 1994; 8:869-884.
132. Oshima J, Steinmann KE, Campisi J et al. Modulation of cell growth, p34^{cdc2} and cyclin A levels by SV-40 large T antigen. Oncogene 1993; 8:2987-2993.
133. Wang H-GH, Draetta G, Moran E. E1A induces phosphorylation of the retinoblastoma protein independently of direct physical association between the E1A and retinoblastoma products. Mol Cell Biol 1991; 11:4253-4265.
134. Adamczewski JP, Gannon JV, Hunt T. Simian virus 40 large T antigen associates with cyclin A and p33^{cdk2}. J Virol 1993; 67:6551-6557.
135. Giordano A, McCall C, Whyte P et al. Human cyclin A and the retinoblastoma protein interact with similar but distinguishable sequences in the adenovirus E1A gene product. Oncogene 1991; 6:481-485.
136. Howe JA, Bayley ST. Effects of Ad5 E1A mutant viruses on the cell cycle in relation to the binding of cellular proteins including the retinoblastoma protein and cyclin A. Virology 1992; 186:15-24.
137. Scheidtmann, KH, Mumby MC, Rundell K et al. Dephosphorylation of simian virus large T antigen and p53 protein by protein phosphatase 2A: inhibition by small T antigen. Mol Cell Biol 1991; 11:1996-2003.
138. Scheidtmann KH, Mumby MC, Virshup DM et al. Protein phosphatase 2A dephosphorylates simian virus 40 large T antigen specifically at residues involved in regulation of DNA-binding activity. J Virol 1991; 65:2098-2101.
139. Ludlow JW. Selective ability of S-phase cell extracts to dephosphorylate SV40 large T-antigen in vitro. Oncogene 1992; 7:1011-1014.
140. Rubinfeld B, Souza B, Albert I et al. Association of the *APC* gene product with β-catenin. Science 1993; 262:1731-1734.
141. Su L-K, Vogelstein B, Kinzler KW. Association of the APC tumor suppressor protein with catenins. Science 1993; 262:1734-1737.
142. Marx J. How *p53* suppresses cell growth. Science 1993; 262:1644-1645.
143. Zambetti GP, Levine AJ. Comparison of the biological activities of

wild-type and mutant p53. FASEB J 1993; 7:855-865.
144. Coppes MJ, Campbell CE, Williams BRG. The role of *WT1* in Wilms tumorigenesis. FASEB J 1993; 7:886-895.
145. Hu Q, Dyson N, Harlow E. The regions of the retinoblastoma protein needed for binding to adenovirus E1A or SV40 large T antigen are common sites for mutations. EMBO J 1990; 9:1147-1155.
146. Wiman KG. The retinoblastoma gene: role in cell cycle control and cell differentiation. FASEB J 1993; 7:841-845.
147. Barbeau D, Charbonneau R, Whalen SG et al. Functional interactions within adenovirus E1A protein complexes. Oncogene 1994; 9:359-373.

CHAPTER 5

Conservation of Tumor Suppressor Function

A. Homologs of Human Tumor Suppressor Genes in Other Organisms

The importance of the roles played by tumor suppressors in fundamental processes of cellular proliferation and differentiation is underscored by their presence in non-humans. Homologs of many of the tumor suppressors first identified in humans have been identified in other mammals and in yeast. The retinoblastoma susceptibility gene, Wilms' tumor gene, *p53*, and *NF1* gene all have homologs in other mammals. The extent of sequence conservation varies between these genes and their respective homologs but is nonetheless considerable.[1-4]

One human tumor suppressor gene, *NF1*, is expressed in all human tissues studied and is similarly expressed in other mammals.[5-7] Further, the tissue-specific and developmentally stage-specific expression of two alternatively spliced *NF1* transcripts is conserved across species.[8] This evidence suggests that the NF1 protein is important in regulating cellular proliferation and differentiation in many if not all mammals. Those observations were extended somewhat when a portion of the *NF1* gene, the sequence homologous to the mammalian GTPase activating protein (GAP), was introduced into yeast deficient for the inhibitor of *ras* proteins (IRA-1 and IRA-2).[9,10] These proteins are themselves homologous to mammalian GAP so this complementation reinforces the putative function of the NF1 protein in the *ras* signal transduction pathway and demonstrates the ability of that function to act in related species.

In order to investigate the importance of the *RB1* gene in cell cycle regulation, transgenic mice have been constructed which lack functional *RB1* genes.[11-13] Those mice were found to be non-viable in addition to displaying multiple developmental abnormalities. Such results confirm the pivotal role played by the *RB1* gene product in mouse as well as human development even though mice do not normally develop retinoblastomas as do humans.

B. UNIQUE TUMOR SUPPRESSOR GENES IN OTHER ORGANISMS

In addition to the homologs of human genes identified in non-humans there have been several unique tumor suppressor genes found in *Drosophila.* At least six unique genes have been identified.[14] Since development in *Drosophila* differs greatly from that of mammals the tumor suppressor genes in *Drosophila* affect presumptive adult tissues in the imaginal discs. Even though the process of embryogenesis in mammals is vastly different from metamorphosis in *Drosophila,* the involvement of these tumor suppressor genes in tumor formation may be analogous to the embryonal tumors observed in humans, i.e. Wilms' tumor, rhabdomyosarcoma, and others.

The first identified and best characterized tumor suppressor in *Drosophila* is the lethal (2) giant larvae (*l(2)gl*) gene.[15] Mutations in this gene give rise to malignant tumors of the brain and imaginal discs.[16] Transcriptional regulation of *l(2)gl* expression is achieved through multiple promoters while translational regulation is mediated by signals encoded in the third exon of the gene.[17] To date there have been no homologs of *l(2)gl* identified in mammals. The *fat* locus of *Drosophila* also encodes a tumor suppressor related to the cadherin gene superfamily. It is interesting to note that other tumor suppressor genes, such as the *DCC* gene involved in human colorectal cancer, encode cell adhesion molecules.[18] It is conceivable that cells become aggressively malignant and dislodge from the primary tumor mass when they become deficient for a particular cell adhesion molecule. Mutations in the *fat* gene lead to hyperplastic overgrowth of imaginal discs, defective morphogenesis, and pupal death.[19]

Another *Drosophila* tumor suppressor gene, the lethal(1)discs large-1 gene (*dlg*) encodes a guanylate kinase homolog most likely involved with signal transduction.[20] Within the *dlg* encoded protein there is a region with homology to the SH3 domain common to nonreceptor tyrosine kinases typified by the protooncogenes *src* and c-*abl,* among others. The *NF1* gene of humans is also involved with signal transduction through its role in mediating the conversion of *ras*-GTP to *ras*-GDP.[21,22] Together these observations demonstrate the importance of signal transduction in regulating cellular proliferation and the sensitivity of that process to perturbations.

The third tumor suppressor of *Drosophila* that affects imaginal disc development is the lethal(2)tumorous imaginal discs (*l(2)tid*) gene. Larvae homozygous for mutations in the *l(2)tid* gene are larger than normal and have a delayed onset of pupariation.[23] The neoplasms forming in these flies have characteristics similar to neoplasms of mammals in that they lose the ability to differentiate.

The *Drosophila* tumor suppressor gene lethal(2)aberrant immune response 8 gene (*air8*) encodes a protein homologous to the human S6 ribosomal protein gene.[24] The phenotype of mutations in this gene includes overgrowth of the larval lymph glands involving both abnormally large cells and hyperproliferation, and underdevelopment of other larval tissues resulting in larval death. The human S6 ribosomal protein is atypical in that it undergoes a developmental stage-specific pattern of phosphorylation and that abnormalities in that pattern are related to cell growth abnormalities and tumor development.

C. SUMMARY

Our understanding of the normal role of tumor suppressor genes in regulating cellular proliferation and differentiation has been enhanced by their study in other organisms. The discovery and study of non-human homologs with strong functional and sequence similarities supports the contention that their roles are important to cellular function. Likewise, the discovery of tumor suppressor genes unique to *Drosophila* suggests that either we have not yet identified their respective homologs in mammals or that they may have been discarded through evolution. In the latter case either their function was supplanted by that of other genes or it became unnecessary because of fundamental changes in development exemplified by the differences between mammalian embryogeny and insect metamorphosis.

REFERENCES

1. Destree OHJ, Kelvin TL, Peterson-Maduro LJ et al. Structure and expression of the *Xenopus* retinoblastoma gene. Dev Biol 1992; 153:141-149.
2. Levine AJ. The p53 tumor suppressor gene and product. Biol Chem 1993; 374:227-235.
3. Bernards A, Snijders AJ, Hannigan GE et al. Mouse neurofibromatosis type 1 cDNA sequence reveals high degree of conservation

of both coding and non-coding mRNA segments. Hum Mol Genet 1993; 2:645-650.
4. Sharma PM, Bowman M, Madden SL et al. RNA editing in the Wilms' tumor susceptibility gene, WT1. Genes & Devel 1994; 8:720-731.
5. Gutmann DH, Wood DL, Collins FS. Identification of the neurofibromatosis type 1 gene product. Proc Natl Acad Sci USA 1991; 88:9658-9662.
6. DeClue JE, Cohen BD, Lowy DR. Identification and characterization of the neurofibromatosis type 1 protein product. Proc Natl Acad Sci USA 1991; 88:9914-9918.
7. Huynh DP, Lin CT, Pulst SM. Expression of neurofibromin, the neurofibromatosis 1 gene product: studies in human neuroblastoma cells and rat brain. Neurosci Lett 1992; 143:233-236.
8. Andersen LB, Ballester R, Marchuk DA et al. A conserved alternative splice in the von Recklinghausen neurofibromatosis (NF1) gene produces two neurofibromin isoforms, both of which have GTPase-activating protein activity. Mol Cell Biol 1993; 13:487-495.
9. Buchberg AM, Cleveland LS, Jenkins NA et al. Sequence homology shared by neurofibromatosis type-1 gene and IRA-1 and IRA-2 negative regulators of the RAS cyclic AMP pathway. Nature 1990; 347:291-294.
10. Ballester R, Marchuk D, Boguski M et al. The NF1 locus encodes a protein functionally related to mammalian GAP and yeast IRA proteins. Cell 1990; 63:851-859.
11. Lee EY, Chang CY, Hu N et al. Mice deficient for Rb are nonviable and show defects in neurogenesis and haematopoiesis. Nature 1992; 359:288-294.
12. Jacks T, Fazeli A, Schmitt EM et al. Effects of an Rb mutation in the mouse. Nature 1992; 359:295-300.
13. Clarke AR, Maandag ER, van Roon M et al. Requirement for a functional Rb-1 gene in murine development. Nature 1992; 359:328-330.
14. Loffler T, Wismar J, Sass H et al. Genetic and molecular analysis of six tumor suppressor genes in *Drosophila melanogaster*. Environ Health Persp 1990; 88:157-161.
15. Mechler BM, McGinnis W, Gehring WJ. Molecular cloning of lethal(2)giant larvae, a recessive oncogene of *Drosophila melanogaster*. EMBO J 1985; 4:1551-1557.
16. Merz R, Schmidt M, Torok I et al. Molecular action of the l(2)gl tumor suppressor gene of *Drosophila melanogaster*. Environ Health Persp 1990; 88:163-167.
17. Strand D, Torok I, Kalmes A et al. Transcriptional and translational regulation of the expression of the l(2)gl tumor suppressor gene of *Drosophila melanogaster*. Adv Enz Regul 1991; 31:339-350.
18. Fearon ER, Cho KR, Nigro JM et al. Identification of a chromosome 18q gene that is altered in colorectal cancers. Science 1990; 247:49-56.
19. Mahoney PA, Weber U, Onofrechuk P, et al. The fat tumor suppressor gene in *Drosophila* encodes a novel member of the cadherin

gene superfamily. Cell 1991; 67:853-868.
20. Woods DF, Bryant PJ. The discs-large tumor suppressor gene of *Drosophila* encodes a guanylate kinase homolog localized at septate junctions. Cell 1991; 66:451-464.
21. Johnson MR, DeClue JE, Felzmann S et al. Neurofibromin can inhibit *ras*-dependent growth by a mechanism independent of its GTPase-accelerating function. Mol Cell Biol 1994; 14:641-645.
22. al-Alawi N, Xu G, Clark R et al. Differential regulation of cellular activities by GTPase-activating protein and NF1. Mol Cell Biol 1993; 13:2497-2503.
23. Kurzik-Dumke U, Phannavong B, Gundacker D et al. Genetic, cytogenetic and developmental analysis of the *Drosophila melanogaster* tumor suppressor gene lethal(2)tumorous imaginal discs (l(2)tid). Differentiation 1992; 51:91-104.
24. Watson KL, Konrad KD, Woods DF et al. *Drosophila* homolog of the human S6 ribosomal protein is required for tumor suppression in the hematopoietic system. Proc Natl Acad Sci USA 1992; 89:11302-11306.

CHAPTER 6

Future Directions for Tumor Suppressor Gene Research

At the present time, molecular studies have identified at least 11 tumor suppressor genes from human tumors. With the burgeoning evidence presented thus far of the importance of tumor suppressor gene function in cell cycle regulation as well as tumor progression, it is not surprising that the search for as yet unidentified tumor suppressors is proceeding at a breakneck pace.

A. REVEALING THE EXISTENCE OF POTENTIAL TUMOR SUPPRESSOR GENES

The observation that deletions of specific chromosomes correlates with specific types of cancers suggests that such tumors may share some common genetic defects. In the quest to identify and functionally analyze tumor suppressor genes involved in a variety of human malignancies, the use of somatic cell hybridization has proved invaluable for localizing their position within the genome. Just such a strategy was recently employed to provide evidence that there is a second tumor suppressor gene, other than p53, located on human chromosome 17.[1] While hybrid cells between a peripheral neuroepithelioma cell line and human fibroblasts were found not to be tumorigenic, hybrids between neuroepithelioma and soft tissue sarcomas remained highly tumorigenic. These results suggested the existence of at least one common genetic defect in the control of the tumorigenic potential of these cells. This defect was subsequently localized to chromosome 17, at a different location than the p53 gene. Fine mapping studies will need to be carried out to discern if this is indeed a new tumor suppressor gene, or one of the already known ones. For example, *NF1* also maps to chromosome 17 and disruption of this gene is common in neuroblastomas, a tumor type derived from cells closely related to the neuroepithelioma cells used in this study.

Along this same line, another hot spot for genetic abnormalities associated with different neoplasms is the short arm of chromosome 3, suggesting that one or more tumor suppressor genes may be located in this region. Following the development of a rapid genetic assay system which permitted functional analysis of defined regions of 3p in the suppression of tumorigenicity in vivo, direct proof defining a tumor suppressor locus at 3p21-p22 was obtained.[2] Using interspecific microcell hybrids containing fragments of chromosome 3p, rapid and unambiguous identification of the donor chromosome, delineation of the role of tumor suppression in vivo, and localization of the locus, was achieved.

B. IMPLICIT TUMOR SUPPRESSOR GENE FUNCTION

Loss of heterozygosity in some human embryonal tumors has implied the existence of a tumor suppressor gene on chromosome 11. *H19*, which maps to 11p15.5, was originally isolated as a gene whose transcription increases upon cellular differentiation and is expressed in a variety of fetal tissues, including the precursor tissues of Wilms' tumor, malignant rhabdoid tumor, and embryonal rhabdomyosarcoma.[3,4] Upon transfection with an *H19* expression construct, two embryonal tumor lines showed growth retardation, morphological changes, loss of growth in soft agar, and loss of tumorigenicity in nude mice.[5] Since *H19* RNA is associated with nonpolysomal proteins in the cytoplasm, this gene may function directly as the RNA component of a ribonucleoprotein.[6] Should this turn out to be the case, then this gene will define a new class of tumor suppressor genes.

C. S-TYPE LECTINS AS POTENTIAL TUMOR SUPPRESSORS

Members belonging to a family of β-galactoside-binding lectins, all of which possess related amino acid sequences, are referred to as S-type or S-Lac lectins.[7] Although lacking an identifiable signal peptide, some members are nevertheless externalized on the cell surface by an atypical mechanism.[8,9] While the precise functions of these lectins are not yet fully appreciated, there is evidence to suggest that one or more of these proteins are involved in growth regulation, cell adhesion, and cell migration. For example, the progression from normal mucosa to adenoma to carcinoma is associated with significant changes in nuclear localization and expression of Mac-2,[10] a 35kDa carbohydrate binding protein (for review see Huges[11]). Those results argue in favor of investigating further the possibility that Mac-2, and perhaps other members of this S-type lectin group, are indeed tumor suppressors.

D. INTERFERONS AS TUMOR SUPPRESSORS

Interferons consist of a family of secreted, multifunctional proteins which serve a defensive role for vertebrates against viral, bacterial, and parasitic infections in addition to certain types of tumors. A recent review by Lengyel,[12] information from which is summarized below, details both direct and indirect evidence that several of these proteins may have tumor suppressor activities.

The double stranded RNA activatable protein kinase (PKR), which is expressed constitutively at low levels in a variety of cell types, can phosphorylate the eukaryotic transcription factor eIF-2. Such phosphorylation sequesters the guanine nucleotide exchange factor eIF2-B thus preventing the recycling of eIF-2-GDP to eIF-2-GTP which will therefore impair protein synthesis. Mechanistically, it has been speculated that this kinase can control peptide chain initiation and, thus, tumorigenicity, since loss of control of peptide initiation due to overexpression of another initiation factor, eIF-4, can do so.[13,14] Alternatively, this enzyme can activate the transcription factor NF-κB by phosphorylation, and thus inactivating, I-κB, its regulatory subunit, in vitro.[15] This transcription factor is involved in the activation of a large number of genes, perhaps including mediators of PKR tumor suppressing activity.[16]

Interferon regulatory factors 1 and 2 (IRF-1 and IRF-2) can modulate the expression of interferons and some of the interferon-inducible proteins. When IRF-2 was overexpressed in mouse cells, the cells became transformed and tumorigenic.[17] The transformed phenotype could be reversed upon overexpression of IRF-1 in these cells. Thus, it is the proportion of these two proteins which appears to exert growth control. Taken one step further, it has been proposed that even loss of a single *IRF-1* allele might afford a significant growth advantage.

Oncoproteins produced by several DNA tumor viruses have the ability to impair interferon action. For example, the adenovirus E1A protein can inhibit gene activation by alpha, beta, and gamma interferons most likely by impairing the formation of an active transcription factor complex.[18,19] EBNA-2, which is involved in the immortalizing ability of Epstein-Barr virus in B cells, also makes the proliferation of such cells resistant to these three interferons.[20-22] Further, EBNA-2 can impair interferon-inducible gene activation.[23]

E. SYNTHETIC TUMOR SUPPRESSORS

Recent advances in our understanding of the normal role of tumor suppressor proteins in regulating cellular proliferation and differentiation, coupled with advances in protein engineering, make the synthesis of "designer" tumor suppressors feasible. Introduction of the genes encoding these proteins by recombinant DNA technologies supports this promise. Specific desirable properties can be designed into these proteins such as increased stability, enhanced function, and regulatable expression. At least one laboratory has reported the synthesis of a growth regulating protein, namely an oncogene which promotes tumor growth based on sequences occurring in the epidermal growth factor, providing a practical demonstration of the applicability of this approach.[24]

F. INSIGHTS INTO NORMAL CELLULAR PROCESSES

Our increasing understanding of the function of tumor suppressor gene products is providing insights into the normal processes which regulate cellular proliferation and differentiation. Different mechanisms which lead to tumor suppressor inactivation are coming to light. The original hypothesis regarding tumor suppressors proposed that tumors may arise as the result of the biallelic inactivation of tumor suppressor genes.[25,26] This inactivation was thought be due solely to alterations of tumor suppressor genes at the DNA level. Indeed many studies have revealed various chromosomal mechanisms which lead to tumor suppressor loss or inactivation.[27-35]

However two recent studies have revealed a role played by RNA editing in the regulation of tumor suppressor gene expression. Sharma and coworkers[36] found that two Wilms' tumor transcripts in rat arose by the process of editing. A single leucine is replaced by a proline in the edited mRNA which reduced the activity of the WT1 protein by approximately 30%. The two transcript forms are expressed in different proportions at different developmental stages suggesting that they play a role in cellular proliferation and differentiation.

RNA editing also affects the *NF1* transcript (see Fig 6.1).[37] The transition of a C to a U results in the introduction of a stop codon at the 5' end of the GAP related domain. RNA editing was observed, at varying levels, in all cell types studied and varied at

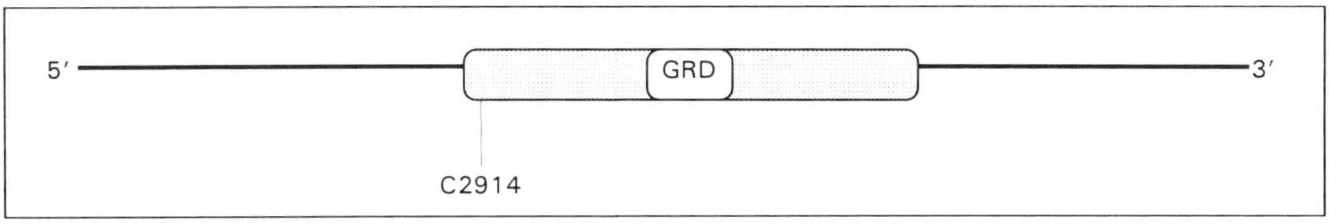

Fig. 6.1. The edited cytosine in the NF1 *mRNA lies at the 5′ end of the GAP related domain (GRD). Editing of the C at that site creates an in-frame stop codon which prevents translation of the only known functional domain of neurofibromin thereby inactivating the tumor suppressor.*

different stages of the cell cycle. These observations support the contention that RNA editing may play a role in regulating proliferation. This most certainly affects the function of the *NF1* encoded protein and may also have an effect on the stability of the *NF1* mRNA. It is interesting to note that inactivation of the *NF1* transcript by RNA editing will have the same effect as biallelic inactivation of the *NF1* gene by some chromosomal mechanism but would not be detected experimentally by methods directed at DNA level alterations. Thus, while the original two-hit hypothesis may be correct in terms of tumor suppressor gene inactivation, the mechanisms leading to that inactivation may be varied.

G. GENE THERAPY

One of the most promising aspects of tumor suppressor genes lies in their involvement with human cancers. Since the role of tumor suppressors is to suppress cellular proliferation, tumors arise when their gene products are inactive or absent. In light of this it is conceivable that introduction of a functional tumor suppressor gene, into a susceptible cell, can ameliorate tumor growth. Many tumor suppressor genes may lend themselves to this therapeutic approach including, but not limited to, the *NF1*, *NF2*, *RB1*, *p53*, and *WT1* genes since these genes have been isolated and their sequences determined.

In vitro evidence exists which supports the potential for tumor suppressor gene restoration. In order to confirm the tumor suppressor activity of the Wilms' tumor gene, Weissman and colleagues introduced a normal human chromosome 11, the one harboring the Wilms' tumor gene, into a Wilms' tumor cell line.[38] Microcell fusion was used to introduce a single t(X;11) human chromosome into tumor cells. The cells receiving the chromosome 11 showed a complete loss of tumorigenicity in nude mice while those receiving an intact X chromosome, or a chromosome 13 carrying a normal *RB1* gene, did not.

Huang and coworkers[39] introduced a recombinant expression vector containing a functional *RB1* gene into osteosarcoma cells and observed a reduction in tumorigenicity both in vitro and in vivo. The parent vector lacking the *RB1* gene did not demonstrate a similar effect, nor did the recombinant vector when introduced into tumor cells which already contained a functional *RB1* gene. In contrast to this report, another group found that restoration of pRB into cells previously devoid of this protein had only a small effect on growth and was insufficient to revert the malignant phenotype.[40] Along this same line, another report has concluded that while pRB can suppress the growth of many tumor cell lines, its growth suppression effect is not tumor cell specific.[41] Taken together, these results suggest that pRB is a relatively weak tumor suppressor and provides the caveat that such non-targeted gene replacement therapy must be approached with caution.

Recently, different levels of pRB were required to rescue two different phenotypes resulting from pRB deficiency; death during murine fetal development and susceptibility to cancer.[42] These findings support another report that the effect of *RB1* on overall mouse development is closely dependent upon its dosage.[43] These results suggest the intriguing possibility that relatively small amounts of pRB expression would be sufficient to prevent tumor formation in cancer-prone cells. Obviously, the formidable task of successfully introducing the gene and achieving stable expression in somatic cells in vivo must be overcome. Even though the obstacles confronting potential gene therapists are formidable and currently not entirely known, tumor suppressor genes represent natural and accessible targets for such therapy.[44]

H. SUMMARY

The field of tumor suppressor gene research is a dynamic and exciting one. On the one hand new discoveries are arising so frequently that an investigator needs to expend considerable effort simply staying abreast of the literature. On the other hand it is becoming amply clear that the proteins encoded by tumor suppressor genes play pivotal roles in regulating cellular differentiation and proliferation. While the original hypothesis that led to the discovery and isolation of the retinoblastoma susceptibility gene was based on the assumption that the product of that gene was a tissue specific growth inhibitor, it was later learned that the retinoblastoma tumor suppressor protein acts in a variety of cell types. The reason, for example, that individuals predisposed to retinoblastoma develop tumors of the retina and not many other tissues is most likely that their other tissues have redundant mechanisms to accommodate a deficiency of the retinoblastoma protein.

Some tumor suppressor genes encode proteins of such great importance that their deficiency has been implicated in a number of tumor types; p53 is one such gene. Another tumor suppressor gene which appears to be involved in many tumor types, including those arising from lung, breast, brain, bone, skin, bladder, kidney, ovary, and lymphocyte, is p16.[45] Perhaps other tumor suppressor genes will be identified which play a role in a greater number of tumors and those genes may provide targets for gene therapy to ameliorate cancer.

REFERENCES:

1. Chen P, Ellmore N, Weissman BE. Functional evidence for a second tumor suppressor gene on human chromosome 17. Mol Cell Biol 1994; 14:534-542.
2. Killary AM, Wolf ME, Giambernardi TA et al. Definition of a tumor suppressor locus within human chromosome 3p21-p22. Proc Natl Acad Sci USA 1992; 89:10877-10881.
3. Davis RL, Weintraub H, Lassar, AB. Expression of a single transfected cDNA converts fibroblasts to myoblasts. Cell 1987; 51:987-1000.
4. Pachnis V, Brannan CI, Tilghman SM. The structure and expression of a novel gene activated in early mouse embryogenesis. EMBO J 1988; 7:673-681.
5. Hao Y, Crenshaw T, Moulton T et al. Tumour-suppressor activity of *H19* RNA. Nature 1993; 365:764-767.
6. Brannan CI, Dees EC, Ingram RS et al. The product of the *H19* gene may function as an RNA. Mol Cell Biol 1990; 10:28-36.
7. Barondes SH, Castronovo V, Cooper DNW et al. Galectins: A family of animal β-galactoside-binding lectins. Cell 1994; 76:597-598.
8. Lindstedt R, Apodaca G, Barondes SH et al. Apical secretion of a cytosolic protein by Madin-Darby Canine Kidney cells. J Biol Chem 1993; 268:11750-11757.
9. Sato S, Burdett I, Huges RC. Secretion of the baby hamster kidney 30kDa galactose-binding lectin from polarized and nonpolarized cells: a pathway independent of the endoplasmic reticulum-Golgi complex. Exp Cell Res 1993; 207:8-18.
10. Lotz MM, Andrews CW Jr, Korzelius CA et al. Decreased expression of Mac-2 (carbohydrate binding protein 35) and loss of its nuclear localization are associated with the neoplastic progression of colon carcinoma. Proc Natl Acad Sci USA 1993; 90:3466-3470.
11. Huges RC. Mac-2: a versatile galactose-binding protein of mammalian tissues. Glycobiol 1994; 4:5-12.
12. Lengyel P. Tumor-suppressor genes: news about the interferon connection. Proc Natl Acad Sci USA 1993; 90:5893-5895.
13. Lazaris-Karatzas A, Montine S, Sonenberg N. Malignant transformation by a eukaryotic initiation factor subunit that binds to mRNA

5' cap. Nature 1990; 345:544-547.
14. De Benedetti A, Rhoades RE. Overexpression of eukaryotic protein synthesis initiation factor 4E in HeLa cells results in aberrant growth morphology. Proc Natl Acad Sci USA 1990; 87:8212-8216.
15. Kumar A, Haque SJ, Hiscott J et al. The interferon induced p68 protein kinase regulates transcription factor NF-κB via phosphorylation of its inhibitor I-κB. J Interferon Res 1992; 12:(Suppl 1)S207.
16. Baeuerle BA. The inducible transcription activator NF-κB: regulation by distinct protein subunits. Biochim Biophys Acta 1991; 1072:63-80.
17. Harada H, Kitagawa M, Tanaka N et al. Anti-oncogenic and oncogenic potentials of interferon regulatory factors-1 and -2. Science 1993; 259:971-974.
18. Kalvakolanu DVR, Bandyopadhyay SK, Harter ML et al. Inhibition of interferon-inducible gene expression by adenovirus E1A proteins: block in transcriptional complex formation. Proc Natl Acad. Sci USA 1991; 88:7459-7463.
19. Gutch MJ, Reich NC. Repression of the interferon signal transduction pathway by the adenovirus *E1A* oncogene. Proc Natl Acad Sci USA 1991; 88:7913-7917.
20. Cohen JI, Wang F, Mannick J et al. Epstein-Barr virus nuclear protein 2 is a key determinant of lymphocyte transformation. Proc Natl Acad Sci USA 1989; 86:9558-9562.
21. Hammerschmidt W, Sugden B. Genetic analysis of immortalizing functions of Epstein-Barr virus in human B lymphocytes. Nature 1989; 340:393-397.
22. Aman P, von Gabain A. An Epstein-Barr virus immortalization associated gene segment interferes specifically with the IFN-induced anti-proliferative response in human B-lymphoid cell lines. EMBO J 1990; 9:147-152.
23. Kanda K, Decker T, Aman P et al. The EBNA2-related resistance towards alpha interferon (IFN-α) in Burkitt's lymphoma cells effects induction of IFN-induced genes but not the activation of transcription factors ISGF-3. Mol Cell Biol 1992; 12:4930-4936.
24. Stern DF, Hare DL, Cecchini MA et al. Construction of a novel oncogene based on synthetic sequences encoding epidermal growth factor. Science 1987; 235:321-324.
25. Knudson AG. Mutation and Cancer: statistical study of retinoblastoma. Proc Natl Acad Sci USA 1971; 68:820-823.
26. Comings DE. A general theory of carcinogenesis. Proc Natl Acad Sci USA 1973; 70:3324-3327.
27. Schwechheimer K, Cavenee WK. Genetics of cancer predisposition and progression. Clin Invest 1993; 71:488-502.
28. Mulligan LM, Matlashewski GJ, Scrable HJ et al. Mechanisms of p53 loss in human sarcomas. Proc Natl Acad Sci USA 1990; 87:5863-5867.
29. James CD, Mikkelsen T, Cavenee WK et al. Molecular genetic aspects of glial tumour evolution. Cancer Surveys 1990; 9:631-644.
30. Scrable HJ, Sapienza C, Cavenee WK. Genetic and epigenetic losses of heterozygosity in cancer predisposition and progression. Adv

Cancer Res 1990; 54:25-62.

31. Cavenee WK. Mitotic abnormalities leading to cancer predisposition and progression. Cancer Detection & Prevention. 1989; 14:269-273.
32. Seemayer TA, Cavenee WK. Molecular mechanisms of oncogenesis. Lab Invest 1989; 60:585-599.
33. James CD, Carlbom E, Nordenskjold M et al. Mitotic recombination of chromosome 17 in astrocytomas. Proc Natl Acad Sci USA 1989; 86:2858-2862.
34. Higgins MJ, Hansen MF, Cavenee WK et al. Molecular detection of chromosomal translocations that disrupt the putative retinoblastoma susceptibility locus. Mol Cell Biol 1989; 9:1-5.
35. Cavenee WK. Chromosomal mechanisms in the initiation of human familial mixed cancers. Internatl Symp Princess Takamatsu Cancer Res Fund 1989; 17:171-179.
36. Sharma PM, Bowman M, Madden SL et al. RNA editing in the Wilms' tumor susceptibility gene, WT1. Genes & Devel 1994; 8:720-731.
37. Skuse GR, Metheny LJ, and Cappione AJ. Editing of the neurofibromatosis type 1 encoded transcript. Submitted for publication.
38. Weissman BE, Saxon PJ, Pasquale SR et al. Introduction of a normal human chromosome 11 into a Wilms' tumor cell line controls its tumorigenic expression. Science 1987; 236:175-180.
39. Huang H-JS, Yee J-K, Shew JY et al. Suppression of the neoplastic phenotype by replacement of the RB gene in human cancer cells. Science 1988; 242:1563-1566.
40. Muncaster MM, Cohen BL, Phillips RA et al. Failure of *RB1* to reverse the malignant phenotype of human tumor cell lines. Cancer Res 1992; 52:654-661.
41. Fung Y-KT, T'Ang A, Murphree AL et al. The Rb gene suppresses the growth of normal cells. Oncogene 1993; 8:2659-2672.
42. Chang C-Y, Riley DJ, Lee EY-HP et al. Quantitative effects of the *retinoblastoma* gene on mouse development and tissue-specific tumorigenesis. Cell Gro and Diff 1993; 4:1057-1064.
43. Bignon Y-J, Chen Y, Chang C-Y et al. Expression of a retinoblastoma transgene results in dwarf mice. Genes & Devel 1993; 7:1654-1662.
44. Friedmann T. Gene therapy of cancer through restoration of tumor-suppressor functions? Cancer 1992; 70:1810-1817.
45. Kamb A, Gruis NA, Weaver-Feldhaus J et al. A cell cycle regulator potentially involved in genesis of many tumor types. Science 1994; 264:436-440.

INDEX

Page numbers in italics denote figures (f) and tables (t).